Cambridge Geographical Series.

GENERAL EDITOR :—F. H. H. GUILLEMARD, M.D.,
LATE LECTURER IN GEOGRAPHY IN THE UNIVERSITY OF CAMBRIDGE.

OUTLINES

OF

MILITARY GEOGRAPHY.

OUTLINES

OF

MILITARY GEOGRAPHY

BY

T. MILLER MAGUIRE, LL.D.

OF THE INNER TEMPLE, BARRISTER-AT-LAW ;
LIEUTENANT, INNS OF COURT RIFLE VOLUNTEERS.

Cambridge:

AT THE UNIVERSITY PRESS.

1899

CAMBRIDGE UNIVERSITY PRESS
Cambridge, New York, Melbourne, Madrid, Cape Town,
Singapore, São Paulo, Delhi, Tokyo, Mexico City

Cambridge University Press
The Edinburgh Building, Cambridge CB2 8RU, UK

Published in the United States of America by
Cambridge University Press, New York

www.cambridge.org
Information on this title: www.cambridge.org/9781107648012

First published 1899
First paperback edition 2011

A catalogue record for this publication is available from the British Library

ISBN 978-1-107-64801-2 Paperback

Additional resources for this publication at www.cambridge.org/9781107648012

Dedicated, by permission,

TO

GENERAL HIS ROYAL HIGHNESS

The Duke of Connaught and Strathearn

(PRINCE ARTHUR),

K.G., P.C., K.T., K.P., G.C.S.I., G.C.M.G., G.C.I.E., K.C.B.,

WHOSE KEEN INTEREST

IN EVERY DETAIL OF THE MILITARY ART

RECEIVES THE HEARTY ADMIRATION

OF EVERY SECTION OF OUR NATIONAL FORCES;

AND WHOSE VARIED EXPERIENCES

IN SO MANY OF THE TERRITORIES BELONGING TO

THE VAST EMPIRE OF HER MAJESTY THE QUEEN

ARE IN THEMSELVES

AN EPITOME OF STRATEGIC GEOGRAPHY.

CONTENTS.

CHAPTER VII.

CHAPTER VIII.

CHAPTER IX.

CHAPTER X.

CHAPTER XI.

CHAPTER XII.

CHAPTER XIII.

CHAPTER XIV.

MAPS AND ILLUSTRATIONS.

Illustrations 1, 7, 17, 19, 22, and 26 are by the kind permission of Field-Marshal Lord Roberts, V.C., Lieut. Vandeleur, D.S.O. Scots Guards, Viscount Fincastle, V.C., 16th Lancers, and Mr E. Wagstaff.

*This map is available for download from www.cambridge.org/9781107648012

CHAPTER I.

THE IMPORTANCE OF THE STUDY OF MILITARY GEOGRAPHY.

THE history of Europe and America for the last fifty years must have shaken the convictions of even the most obstinate among those who hold that the regeneration of society is to be accomplished by peace, and by peace alone.

No historian for some generations to come will venture to assert, as did the late Rev. R. Green, that war plays a small part in the real story of European nations. Wars have of late been frequent on both sides of the Atlantic, and have been waged with a destructive fury far beyond the imaginations of even Napoleonic warriors.

In the United States of America, a civil struggle, starting with abstract principles of philanthropy and mere constitutional quibbles, developed into a conflict which in four years caused the Federals alone a loss of 500,000 men, and about £1,000,000,000 of money. Nor was there, as some believe, any mitigation of the horrors of war; the mine at Petersburg recalled the breach of Badajoz; Grant at Cold Harbour rivalled the pertinacious assaults of Napoleon at Borodino; the passages of the Rappahannock were as deadly as any battles on the Elbe; starvation was employed as ruthlessly at Richmond and at Vicksburg as at Londonderry or at Ulm.

Reille or Suchet would not have dared to make such a devastating clearance of the property and people of Navarre or Valencia as was effected in Georgia by Sherman. Sheridan proved that the stern exigencies of conquest applied as fully to the beautiful valley of the Shenandoah in the nineteenth century as to Saxony in the seventeenth, or to Bavaria in the eighteenth.

The generation of young Britons who lived between the date of the campaign of Waterloo and that of the Crimea, not having before their own eyes any spectacle of the vastness of war, studied with amazement records of the old migrations of warriors, who, under the wild leadership of Gothic or Hunnish chiefs, imposed the barbarism of Asia on the art and opulence of southern and western Europe. But the Franco-German war began August 2nd, 1870, and before the 1st of October all Gibbon's records of invasion had ceased to surprise. Far different were the intellectual culture, the personal habits, and the moral aims of Attila and Von Moltke, but as far as Gaul was concerned the result was the same, the regular military forces disappeared by hundreds of thousands, dead or prisoners, at the end of one month; the citizens of the vast and splendid capital were shut up to starve before the end of the second. Nor could the brilliant efforts of Gambetta and his heroes, nor the dreadful sufferings endured by his hordes of levies, bring about any such retribution as Theodoric and Aetius inflicted on the "Scourge of God." D'Aurelle de Paladines and Faidherbe, Chanzy and Bourbaki, vainly struggling too late, only added to the heaps of slaughter and the prodigious cost. A million of German soldiers crossed the Rhine before the end of 1870, and their successors have ever since remained in Alsace and Lorraine. It is true the invading armies did not return with such loot as Attila brought from Italy to his villa on the northern bank of the Danube, but their leaders exacted no less than £26 from every family in France. How was this? Simply because for a generation the leaders of the German race had devoted themselves with constant zeal to profound studies of strategy and military geography in all its bearings, and during the struggle had acted on the knowledge thus acquired. They were ready for war in its most highly developed form, their adversaries were neither as well trained nor as well prepared. There was no other reason.

But this is only that old oracle of time, hearkening to which gave Philip of Macedon and his "godlike son" the command of Europe and of Asia from Attica to the Oxus—the oracle which was voiced again in the days of Elizabeth by a sage as wise as was the tutor of Alexander. The French and British had forgotten for a time, but the Germans had remembered, Bacon's philosophy:—"above all for Empire and greatness it importeth most that a nation do profess arms as their principal honour, study, and occupation."

In the days of Queen Elizabeth the glory of the Sultan had not yet decayed. For a long time after the death of Bacon, it might be said in his words "by their profession of arms the Turks do wonders." The Greeks in 1897 declared war against the comparatively feeble

successor of Suleyman the Magnificent with hearts as
light as M. Olivier's in 1870. But the Turks in their
political decay had remained soldiers. Greece would
have been even more completely overwhelmed in 1897
than was France in 1870, and the Athenians would
have experienced the fate of the Parisians, but for the
intervention of Europe at large.

Yet the ancestors of these same Greeks had spirits
as stout and warlike as ever inhabited the breasts of
the followers of the Prophet. The small force which
Alexander conducted from Thrace to Syria, and from
Egypt to Bactria, and back through Persia to Babylon
in eleven years, taught posterity this lesson, "walled
towns, stored arsenals and armouries, goodly races of
horse, chariots of war, elephants and ordnance, and the
like,—all this is but sheep in a lion's skin unless the
breed and disposition of the people be stout and warlike."

The fate of Asia two thousand years ago will most
assuredly be the fate of any modern state which is not
ready on any just occasion for arming. Politicians who
are so devoted to the ephemeral questions of the hour
as to neglect a study of the military resources of their
community as compared with those of others, and who
do not organize the power of their state betimes, are
courting disgrace for themselves and disaster for their
nation. The lessons of antiquity have been rewritten in
the largest letters in the annals of our own time; it
behoves all who govern or guide their fellow-citizens to
study these lessons well, and, taught by the dread experiences of others, to be wise in time. If it be clear
that their people are not fit to fight, bold to fight, and

armed at all points for a fight, they should forthwith make all other political and social questions subordinate to the duty of introducing such ordinances, institutions, and customs in this respect as may insure greatness to posterity.

But to turn from strategy by land to strategy by sea —to the British nation a branch of learning which it is now impossible to ignore. A citizen who is quite ignorant of our strategic geography so far as it relates to command of the sea can scarcely comprehend the contents of his newspaper, and is unfit to discuss any economic question relating to the conditions of our national life. The past story of our isles was bound up with the sea; our present existence, our future position depend upon the sea, and there can be no excuse now-a-days for not understanding the whole mystery of Sea Power.

A generation ago the corn of naval strategy was hidden in the chaff of so many bushels of tactical narratives that few laymen could understand how sailors fought battles, or what were the principles underlying long-continued naval manœuvres, either in the Mediterranean Sea, the "narrow seas," or the ocean at large. But the American, Captain Mahan, assisted recently by many Britons, has so cleared up the mystery that the leading features of naval operations, their objects, and the necessary consequences of defeat under certain conditions, ought to be as obvious to every educated man in our isles as the postulates of Euclid.

Such is the supreme importance of the issue,—our retention, or our loss of Sea Power,—that a short course of training in the records of our maritime life should be part of every scholastic curriculum, and a few works like

Mahan's should undoubtedly be included in every University course. If we suffered a series of defeats at sea similar to the battles of Woerth, Spicheren, Vionville and Gravelotte in one fortnight, there can be no doubt that we should have to capitulate at discretion. The enemy need not trouble about a Sedan. But France, in spite of all these, and the fall of Strasburg and Metz, was able to combat with vigour till February, 1871. All classical students are well aware of the value of sea power to the Greeks in their operations against the Asiatics, and when fighting among themselves. The Romans, whose military instinct was for many centuries unerring, determined to secure it, and it was a valuable weapon against Hannibal. The battle of Actium decided the empire of the world; the battle of Lepanto arrested the progress of the Turk.

The old story, like all other strategic stories, was repeated and emphasised by the Elizabethan writers. When Spain was at the height of her power, it became clear to our people that the command of the sea is an "epitome of monarchy," and that the command of both Indies was incidental to command of the sea. American naval experts have clearly apprehended these doctrines ; and the superiority of the American to the Spanish fleet in the recent war (1898) has given the United States control of both Cuba and the Philippines. Our own navy had taken both Havana and Manila after the battles of Ushant and Lagos had made Great Britain ruler of the sea during the Seven Years' War.

The importance, therefore, of the study of strategy and of geography to statesmen is unquestionable. Many

of our international difficulties could never have arisen
had the leaders of either political party been possessed
of any sound knowledge of military or political geography. Our diplomacy has drifted into strange blunders,
not so much through lack of skill as through lack of
knowledge. Delagoa Bay, the key of South Africa, is a
striking example ; so are several districts on the borders
of Canada. The frequent assertions that no Power
could move from the Caspian Sea to the frontiers of
Afghanistan in less than a generation were readily
accepted as political arguments twenty years ago,
although Alexander the Great had traversed the same
territory with ease in a couple of years. The Russians
conquered the Caucasus in 1859, were in Khiva in 1873,
and in Merv in 1884. A very rudimentary knowledge
of the movements of armies from the valleys of the
Euphrates and Oxus, and from the Khanates of Tartary
towards India, from the days of the early Persians to
those of Nadir Shah, would have prevented much wild
talk on public platforms between the years 1876 and
1882.

A popular fallacy much applauded in political
speeches by people who would have known better had
they been taught the rudiments of strategy, is that if any
foreign Power dared to violate the sanctity of our shores,
"our people would rise as one man," and drive the
invaders into the sea. This pious opinion has been
publicly promulgated by more than one statesman of
eloquence and repute. Nothing can be more certain than that not once in history has this kind of
event taken place. Our ancestors were not so foolish ;
they did not wait for Napoleon's invasion to commence

drill; they began to organize and drill before Napoleon accepted war in 1803 and more than a year before he could have possibly arrived, and yet Sir John Moore told Mr Pitt that he dared not bring his volunteers into the first line. The enormous levies of Gambetta, whether on the Somme or the Loire, the Huisne or the Lisaine, —though sometimes four to one and well armed—were repulsed with ease on the offensive or defensive by every German corps commander who came across them. Neither the bravery nor the mountains of the Tyrolese could keep out the French in 1809. The Spanish insurgents, though fanatical, patriotic, and numerous to a remarkable degree, and with every topographical advantage in their favour, whether in the open field as at Ocano, or behind walls as at Saragossa, were crushed, and Spain was only saved by British regular troops. How much misery would have been spared if the ordinary member of a Cabinet only understood that the supply of food to an army, and an efficient ambulance service, are fully as necessary as rifles or cannon. He would learn this in half-an-hour in any strategy class, but the statesmen of England who despatched troops to perish of want in the Crimea, and the wire-pullers of the United States who sent out their soldiers to unhonoured graves in Cuba, were alike lacking in knowledge. In each case the injury caused by the activity of the enemy was small compared with that inflicted by the ignorance of the authorities.

But enough for the general utility of this study. That soldiers of all classes, regular and auxiliary, should be as well versed in strategic geography as opportunities allow, would seem at first sight a self-

evident proposition. Barristers study every precedent
relating to their profession, so do physicians, and so
would the most practical soldier, if he could get time.
"Our officers were not bothered with strategy and
tactics in the days of Marlborough and Wellington," say
some! Yet as a matter of fact young Churchill learned
Vegetius by heart, and Wellington went through careful
studies under a military tutor in France as a boy, and
was a habitual student when a man. When he commanded the army of occupation in France after Waterloo
he regularly studied four hours daily, and declared
that this had been his custom throughout his Indian
career, yet he was also able to find time to be a rider to
hounds. Not only so, but some of the most able soldiers
have also been writers on Military History, from Caesar
to Marmont, and from the Archduke Charles to Marshals
Wolseley and Roberts. The most eminent officers who
led armies to victory during the American Civil War had
gone through a sound course of study in strategic
geography, and all sciences relating to the art of war, at
West Point, one of the best educational institutions in
the world. Jackson was a Professor before he became a
General. Sherman was one of the ablest commanders
on the Federal side, and he felt so keenly the necessity
for a knowledge of geography that he wrote to his friend
Ewing in 1844:—"Every day I feel more and more in
need of an atlas, as the knowledge of geography in its
minutest details is essential to a true military education.
I wish, therefore, you would procure me the best geography and atlas extant." Twenty years later his well-
stored brain enabled him to start with confidence on his
adventurous martial pilgrimages from Chattanooga to

Atlanta, and thence to Savannah by the sea. Yet Sherman was also a great sportsman.

Napoleon during his operations by the banks of the Danube, in 1809, and the banks of the Elbe, in 1813, worked out the campaigns of Marshal Saxe and of Gustavus Adolphus. Before his campaign of 1796 in Italy, the young general Bonaparte procured and studied diligently at considerable cost the best treatises on the geography of northern Italy, and before his campaign of 1815 in Belgium the Emperor Napoleon wrote to his Minister of War :—" Get me a *précis* of everything that has taken place in the past with regard to campaigns along our Eastern frontier, and also state the positions taken up in order to secure co-operation by the armies of the Moselle and the Rhine." Old Marshal Blucher said to his Chief of the Staff—"the brain of the army"—Gneisenau, during the Seine and Marne campaign of 1814, "Gneisenau, if I had only studied, what a man I might have been !" But, though not learned himself, he recognised and used the knowledge of others.

The Prussian Staff Officers, having learned the folly of being behind the times by the disasters of 1806, have devoted themselves to elaborate studies of strategy and geography, not only by means of books, but by travel, and following in person each campaign. For example, Moltke voyaged in Turkey in 1829, and thence to the valley of the Tigris, about which he wrote a very interesting treatise, which Lord Wolseley says he read in the trenches before Sebastopol. German officers have served in Spain, in the Caucasus, in India, in the United States. The superior officers and generals of

Germany are splendidly educated, and of high intellectual capacity.

War never leaves a country as it found it; whether a
nation fails or wins, its whole future is profoundly
modified. No pains are too great, no expense too heavy,
no strain too severe, if the result be success when a
decisive conflict comes, as come it must. The Germans
are the most highly educated nation in the world, and
yet they are a "nation in arms": and their leaders are as
sound in brain as in body, even as was old Moltke when
he directed the passage of the Reisengebirge in 1866, or
of the Vosges in 1870. They are as well versed in pre-
cedents as the ablest judge. They cannot be promoted
until their knowledge and energy and prescience are
beyond dispute. It is equally beyond dispute that
before 1870 the officers of the French army were not
properly qualified to lead any troops. They were
brave and knew the tactics of the parade ground, and
had had some fighting experience in Algeria, but they
did not know the geography of north-eastern France.
Many of the generals had never opened the pages of
modern strategists. In this respect they were far below
Austrian and Russian soldiers, not to speak of their own
immediate foes. They repudiated the notion of organization and of mobilization and schemes carefully drawn
up well in advance. In consequence, after the reverses
of August 6th, they were quite unable to form a plan
that would suit the defence of the Moselle, nor had they
any idea of how to seize and fortify a flank position like
Osman's at Plevna. When at last they tried to recombine
their scattered corps, and MacMahon moved to the
relief of Bazaine, they puzzled their adversaries by the

mere absurdity of their operations. The German staff could not credit the news that MacMahon was marching from Chalons to Metz, close to the neutral frontier of Belgium, and around the right flank of two armies nearly twice as numerous as his own. But once they realised the truth, such was their knowledge of the strategy that would suit the country, that in less than twenty-four hours they had changed the direction of the marches of 250,000 men from east to west, the left leading, to from south to north, the right leading, and proceeded to traverse the roads of the Ardennes as coolly as if they were in Bavaria or Saxony. In a week they had captured MacMahon and his 83,000 exhausted and half-starved followers.

Continuous labour to perfect our national offensive and defensive naval and military systems must, therefore, for a long time to come, be the highest political wisdom. Inasmuch as no veterans among the inhabitants of our islands have had any practical experience of what the consequences of defeat in a decisive campaign and invasion could mean to the United Kingdom, our people are the least prepared for them. The people of the United States are much better prepared ; their older citizens by the banks of the Potomac, the Mississippi, and the James have a lively recollection of the proceedings of advancing and retreating hosts. Having regard not only to the enormous expenditure on armies and navies in Europe, but to the inevitable friction that must follow European expansion in Asia and Africa, every citizen should spend some time in getting a clear conception of the art of war. Machiavelli laid it down in his great masterpiece that a prince is not fit to rule who does not know war. As our people now-a-days govern themselves, they need this knowledge for themselves. As Goltz declares, "They ought to know how to forge weapons, to strengthen their arms in order to carry them, and to steel their hearts so as to endure the hardships which a struggle for the Fatherland entails." Clausewitz says, "The waging of a war is in itself very difficult, of that there can be no doubt ; but the difficulty does not lie alone in the fact that special erudition or great genius is demanded in order to perceive the true principle for conducting war ; of this every well-organised head, who is without prejudice, and who is not utterly ignorant of the matter, is capable. Even the application of these principles upon the map and paper entails no difficulty, and to have sketched out a good *plan of operations* is no great master-piece; the whole difficulty consists in faithfully carrying out the principles one has proposed to oneself."

But apart from their necessary connection with all political and military enterprises on a large scale, narratives of campaigns must ever attract the attention of the intelligent and the curious. Dr Arnold points out that in all ages and among all peoples descriptions of the operations of warriors are most popular, and this not because of any inhuman delight in details of carnage and destruction, but because in battles the highest faculties of the race are exerted in their most intense energy. The struggles of embattled men are perennially interesting to all men, and the history of mankind is the history of armies.

CHAPTER II.

THE LEADING PRINCIPLES OF STRATEGY AND MILITARY GEOGRAPHY.

READERS of Austin's *Jurisprudence* will remember how severe a task he set himself when he attempted to define such simple phrases as "Law in general" and "Municipal Law in particular," and, after all, his definitions leave on the mind a sense of vagueness. To define "Strategy" and to distinguish between Strategy and Tactics are almost as difficult undertakings. The general plan of a campaign is strategy, the details of a battle are tactics. No doubt in some cases the distinction is clear. Marshal Marmont's manœuvres up to the battle of Salamanca can be discriminated exactly from those in the battle itself, but it is hard to separate the action of the three German armies during August 6th, the date of the battles of Spicheren and Woerth, and to say how far the action of the Bavarians on the one side, and the third corps on the other belong to the domain of strategy and how far to tactics. When the Prussians arranged their invasion of Bohemia in 1866, they designed and executed a fine strategic plan, which was in practical operation days before the battle of Sadowa. In fact, when their left army reached Josephstadt, and their right armies drove Clam Gallas from the Iser, Benedek was already beaten from a strategic point of view. He stood to fight on the Bistritz, but he was beaten not by any merely tactical operations, but principally because at a critical moment of the battle the Crown Prince, continuing his original line of march, came into a strategical position on his right flank, and he was thus obliged to retreat or be cut off from the Elbe. He was quite able as far as tactics were concerned to stop the enemy coming against his front from the west, in spite of their breechloaders.

There are certain conditions appertaining to the movements of armies, a proper appreciation of which may lead to a satisfactory definition or description of strategy. Soldiers, like all other men, require a certain minimum of food, sleep, shelter from the weather, and clothing. It is true that at times, and for short periods, generals may call upon their men for rare self-denial; rations may be scarce, and bivouacs take the place of encampments or cantonments, but these periods of deprivation must be slight, or a patient and well-disciplined army will perish and a less orderly army will mutiny. The waggons carrying the supplies must be constantly replenished from the base, or from depôts provided in the theatre of operations. Moreover, many men get ill even during a short campaign : the proportion of sick in King Joseph's army in Spain in 1812 was 51,345 out of a total of 291,379 men all told. Wellington was most particular about the health of his troops, and did his best to provide medical comforts, and his men were recruited from hardy races and agricultural populations, yet he counted 16,984 sick out of about 72,000 on his rolls in 1812. To take the place of men thus invalided and of the dead, a constant stream of recruits flows from the rear to the front after every skirmish or battle. It is also now-a-days usual to send all convalescents home at once by rail. Moreover ammunition is quickly exhausted, and the work of replenishing the magazines causes a circulation of waggons backwards and forwards. The further the army marches the greater become the difficulties; "it drags at each remove a lengthening chain." In some of the semi-barren districts of Europe the difficulties of subsisting an army become terrible. When the Russians reached the environs of Constantinople in 1876, of their enormous hosts not 50,000 could stand to arms. By the end of the first week after he crossed the Niemen, Napoleon had lost 10,000 horses, and this before he had had any serious fighting with the Russians. If troops accustomed to live in cottages sleep during bad weather in the open air they become diseased with astounding rapidity. So dangerous is this that von Werder's troops, during the exciting fights on the Lisaine in January, 1871, were marched some miles daily from the battles so that they might sleep under roofs. When a German division reinforcing Napoleon at Wilna, in 1812, bivouacked one night in the open, the snow covered 6000 corpses before the dawn. Perchance the manœuvres in Wiltshire in 1898 may have given some readers an idea of the difficulties of supplying armies. The troops of the Duke of Connaught and Sir Redvers Buller numbered each about a Napoleonic or German army corps. Therefore only two corps operated, and

they moved from separate bases, north and south, and
had an elaborate railway system available on both sides.
The ablest caterers in England, with ample credit and
exhaustless means, were employed, and yet these small
armies had often to wait for their meals, and thirst
harassed men and horses from start to finish.   Another
lesson was supplied by the state of the roads, crowded
in all directions by teams and waggons of every description.   Fortunately the distances to be traversed from
day to day were only a few miles ; had the distances
been long, the nocturnal camps of famished men would
have been scenes of wild riot, and the peaceful citizens
of the south of England, which has been free from invading footsteps for years, would have seen their own
defenders illustrating some of the horrors of war.

An army of invasion is followed by immense trains.
Edward III's army in France had 6000 waggons, stretch-
ing two leagues.   The combatant part of an English
army corps going by rail would require 104 trains
of about 30 carriages each ; the luggage would require
61 trains more.   If an army corps of 30,000 men and
10,000 horses rests for a day or two preparatory to
a battle or during a siege, it eats up all provisions pro-
curable in a piece of rich country nine miles long by
five miles wide.

But even supposing that food and drink are plentiful
and sleeping-places not too far away from the lines of
march other serious questions arise.   A British division
on the march along an ordinary main road without an
advanced guard would be five miles in length.   Napoleon invaded Belgium (1815) with a small army, as
continental armies are now counted, viz. 125,000 men,

yet his force would have stretched 49 miles on a single road. If the modern German army were put in motion on one road, when the head of the column was marching into Mayence the last company would be at Eydtkuhnen, on the Russian frontier. The whole military road from the Rhine to the Russian frontier would be thickly crowded with soldiers, guns, and transports. If these were to pass out through a single gateway, day and night, it would take a fortnight for all to pass through. If tents were also carried, as was the case in Wiltshire, the length of the column would be considerably increased. Wellington reported to his Government after the battle of Vitoria that he had taken 151 guns and 415 ammunition waggons. The rewards were generous, but few members of Parliament at the time had a clear conception of the enormous bulk of this booty; it would crowd the whole road from Westminster to Woolwich.

Admirable arrangements are made by the Germans for quenching thirst while on the march, for in a long march and in hot weather this is often most trying, but if beverages were also carried to a considerable extent the length of the column would be proportionately increased.

The question of quarters now arises. The Austrian army of 1866 required almost the whole of Moravia for quarters. If modern armies were in line of battle and not in column of march the result would be equally startling. The French army would reach from Épinal to Verdun, a distance of 80 miles, even though the individual regiments were ranged together closely.

From these considerations it is manifest that the

various corps of an army must move not by one, but by many roads to a common object. In some countries these roads may be near to one another, in others they are more distant. In some countries they are separated from each other by rivers, only passable at a few fords or bridges; or worse, by ranges of hills only to be traversed by a few passes.

Any considerations of strategy that apply to ordinary roads apply to railroads with greater force; they facilitate supply and movement tenfold. The German army at Paris, for instance, was supplied for a considerable period (September to December) by one line, which enabled sixteen trains a day to be brought close up to the lines of investment. A railway may be very easily rendered useless by the destruction of a tunnel or of a long via-duct, and therefore requires more care than a road, as its repair would in either case be very tedious and costly. It is well worth the while of an enemy to make a raid against points like these, and the defensive commander could not do better than to organise operations against the line used by an invader. Nothing could more embarrass his opponents; indeed a good scheme for checking a hostile advance against a capital would be the establishment of a place of arms and assemblement on a line perpendicular to the main railroad from the frontier to the capital, and not very far away; a few days' distance would suffice.

As an example of the relative efficiency of the various means of transport the following facts may be instanced :—

One train of 25 to 30 carriages will convey 300 tons of supplies 200 miles in a day. One hundred and fifty large civilian waggons will take the same load 12 to 14 miles in a day. A barge 20 feet long, 6 feet wide, and 4 feet deep will convey a load of nearly 10 tons, so it would require 30 to take the 300 tons of supplies. But it is not necessary to discuss further the superiority of railways over other avenues of supply from the point of view of efficiency and rapidity.

A large army, then, requires many and good roads; a bad road, especially after severe weather, becomes almost impassable if once a division has marched over it. There are many cases recorded in which armies could not move at all because of the inferior quality of the roads, and other cases in which a few heavy showers have thwarted the schemes of strategists. A heavy fall of snow will stop an army. Once the main road from the front of the army to its base, or from the depôts of an army to its sources of supply is blocked a catastrophe is imminent; anxiety pervades all ranks, the pressure of hunger is soon felt, and fear turns into panic and disgrace. Such was the terror which overcame the French when Graham seized their road to Bayonne during the battle of Vitoria; and the panic which overcame the Russians and caused the disgraceful rush to the bridge of Simnitza on July 31st, 1877. In our native wars our troops have often to construct roads as they advance, or, as is especially the case in West Africa, to proceed in single file by jungle paths.

Once the reader clearly understands that soldiering and fighting are far from synonymous—that in a campaign combats are occasional while marching is constant —that before entering into a battle a general must be most careful to secure his line or lines of retreat; he

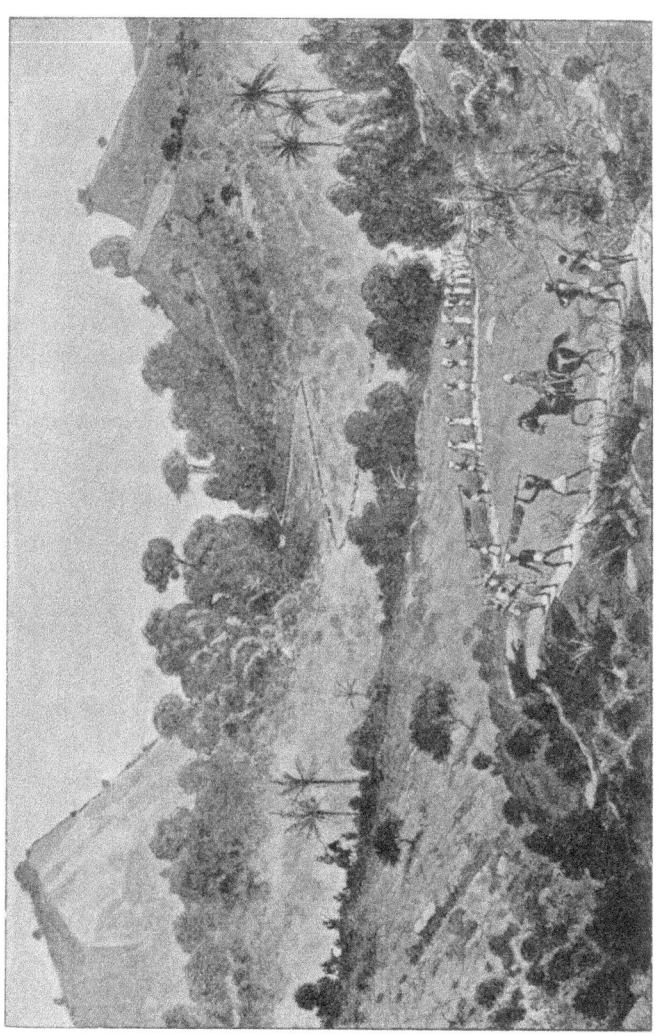

Niger Company's Troops marching to Kabba.

understands the leading principles of strategy, whether
he can define the phrase to his satisfaction or not. He
sees that a general whose road homeward or to his base
is threatened or cut by a superior force must, if he loses
a decisive battle, be ruined as well as defeated ; while a
general who has secured his line of communication will
not be ruined even if defeated, but can fall back, procure
recruits, replenish his waggons, and begin to fight again
with a fair prospect of success.

In order to render success certain, we ought to
choose a line of operations which will enable us to keep
our troops together as much as possible. In order to
enhance our success, we ought to select a line of operations which will bring us as close as possible to the
enemy's line of retreat. But how close, or how far
away, is a delicate point best decided on the spot.
Hood went too far away in 1864, when he marched
to the Tennessee, for he enabled Sherman to do as he
pleased in Georgia.

In order to protect ourselves against the consequences
of a possible defeat, or, at least, to diminish its gravity,
we ought to select such a line of operations as will be as
near as possible to our own line of retreat. Radetsky
had a better chance in 1849 than the Sardinian army,
Marmont a better one than Wellington in 1812.

The relative efficacy and safety of these methods of
attack may be illustrated by the following figure :—

    1. Let us suppose that the enemy is at AB fronting
towards F, and with the line of retreat CD. If we are
at LM on the line of operations DC, which is also the
enemy's line of retreat, our success will be greatest, but
it is less certain, because of the difficulty of concentrating

our forces, and in case of reverse the consequences will
be much more serious.

2.   If we are placed at *IJ* on the line of operations
EF, which is our line of retreat, the operation gives the
best opportunity of concentrating our forces, and the
greatest security in the case of a defeat, but the least
chance of a very considerable success.

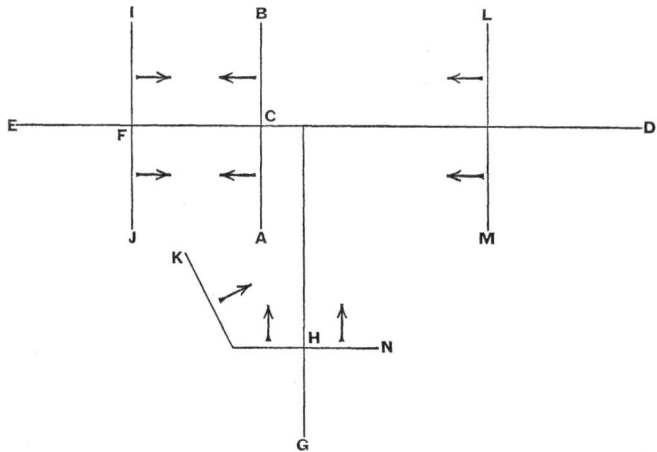

3.   Let us place ourselves at *KN* on the line of
operations *GH* which serves as our line of retreat.   This
operation gives almost as good an opportunity of concentrating our forces, and consequently almost as much
certainty of success, as in the former case; it affords
equal security in case of defeat, and it presents many
more chances of obtaining a decisive victory.

It is not enough to have compelled the enemy to
retreat in a disadvantageous direction, it is also necessary

by our pursuit to keep him in that direction as long as possible; but it is not always that an eccentric retreat is disadvantageous; a retreat that after reorganization threatens the enemy's flank is better than a retreat to the original base. Suppose for example that the French after Woerth had retreated to Langres.

Whatever the plans and preparations for any military enterprise, the result depends on a decisive battle; our own success, or at least the diminution of the enemy's success in a campaign, are determined by the battle. All Marmont's elaborate plans were spoiled by Wellington's success at the action of Salamanca. Had the Germans been beaten at Gravelotte their position would have been very hazardous.

The plan of every enterprise ought to be settled in advance. This is an indispensable condition for arriving at a determined end, but besides the end to be attained the plan should have regard to the nature of the means and to existing circumstances. But there must not be any rigidity in the plan, and the staff must be ever ready for modification and variation during the development of the campaign. Napoleon made three mistakes in his plans in 1806, but the result was not seriously affected thereby. Rustow follows up these ideas very fully in his *Stratégie*, but enough has been said for our purposes.

Before defining accurately strategy and the leading ideas in the mind of a strategist, it is well to direct attention to the fact that inasmuch as an army advances along several roads its various portions may be obliged to move at some distance from each other, and if they cannot keep touch on the left and right flanks respec-

tively they may become so separated for a few days as
to be practically two distinct armies, while if they have
started from different bases they will develop eccentric
or divergent lines of communication.   In this case all
the advantages of their superior numbers may be lost
if the least hitch in arrangement should occur, and they
may be beaten in detail.   For example, if three armies,
each of 100,000 men, were converging towards a given
point from three places, each forty miles apart from the
other, if they did not keep touch for, say, any two days
during a ten days' march towards their object, they
might all three be broken to pieces, driven away from
each other, and excluded from the arena of operations
by an army, each portion of which was linked to the
other, and yet only 200,000 in all.   This is obvious, but
worse still, two of them might be stopped at defiles
or strong positions, while the third was being destroyed
only twenty miles away.   By superior skill, therefore,
and taking advantage of that separation of the masses
of the foe which is sometimes unavoidable for topographical reasons and sometimes accidental, an army of
relatively very inferior numbers may ruin much more
numerous bodies of adversaries.

The length of the line of an invader's communication,
and the fact that in due time his army may break up
into sections, especially if the invaded country be large,
affords chances of victory to resolute patriots who will
not yield an inch of territory, and who are endued with
courage never to submit or yield, or "what is more,
not to be overcome."   It was by acting on these calculations, the accuracy of which admits of mathema-
tical demonstration, that the Archduke Charles, after a

retreat of 200 miles, delivered South Germany from
Jourdan and Moreau in 1796; and that Jackson and
Lee perplexed, harassed, and defeated all the generals
of the North for two consecutive years in the valley of

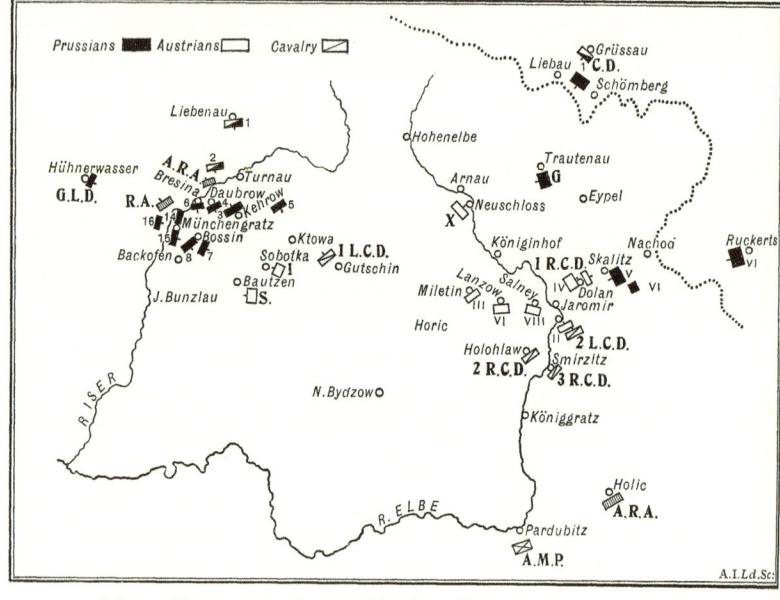

Map to illustrate Armies marching from Divergent Bases on a common
centre and the disposition of Detaining Forces. Prussians at Trautenau and
Nachod detained by Austrian Tenth and Sixth Corps. Prussians marching
to the Iser turn the Austrian detaining force of the 1st Corps and Saxons,
June 28th, 1866.

Hühnerwasser to Trautenau, 40 miles; Trautenau to Pardubitz, 40 miles.

the Shenandoah and along the woods and rivers of
Eastern Virginia.

The Confederate Jackson was at Staunton May 6th,
1862, he fought the Federals Milroy and Schenck near

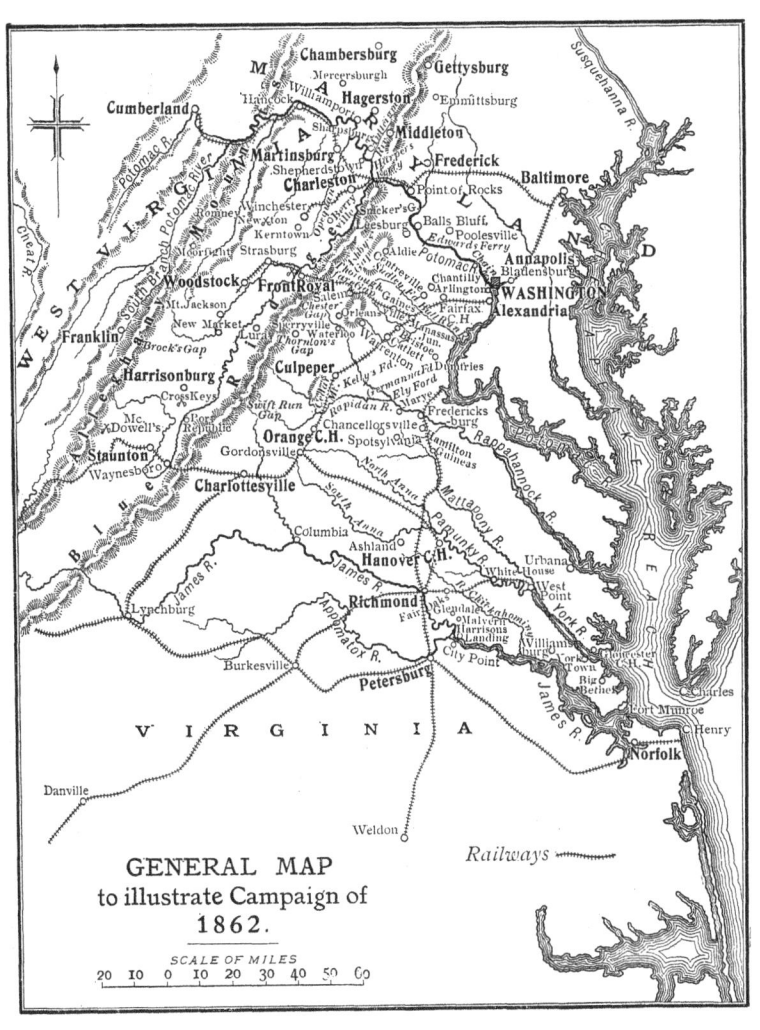

GENERAL MAP
to illustrate Campaign of
1862.

SCALE OF MILES
20 10   0   10 20 30 40 50 60

GENERAL MAP TO ILLUSTRATE THE CAMPAIGN IN VIRGINIA IN 1862.

McDowell's on May 8th and drove them over the hills ; fought Banks at Front Royal, Newtown (near Kerntown), and Winchester May 23rd to 25th, and drove him over the Potomac. He defeated Fremont at Cross Keys June 8th, and Tyler at Port Republic June 9th, and then acting on interior lines went to Richmond, and joining Lee, drove the Federal McClellan from the Chickahominy to the James river, June 26th to July 2nd. In each case he left a detaining force between the portions of the beaten armies.

Bearing in mind the data upon which each is based, the reader may accept Hamley's definitions that "the theatre of war is the province of Strategy, the field of battle is the province of Tactics. The object of Strategy is so to direct the movements of an army that, when decisive collisions occur, it shall encounter the enemy with increased relative advantage." We may adopt the view of Bulow that when the first shot is fired the duties of a strategist are suspended until the close of the fight, when they are resumed. Of course the best laid schemes of a strategist are wasted if he loses a battle, and indeed, if the adversary be so very strong that he can win every battle he can afford to view with some indifference all efforts to cut his line or break up his army into disconnected fragments; but even then he cannot afford to have his road (or his railway) for supply blocked or destroyed ; the resulting injury, material and moral, would be too severe. Accordingly, in 1870—71, the Germans left the *etappen* (or line of communication troops) along the railway from the Rhine to Paris, and when any serious movement of Gambetta's levies against that line was threatened they modified their whole

arrangements forthwith, although invariably successful
in battle against any superiority of numbers.   The vast
army of Russia in 1877 did not dare to cross the Balkans
so long as Osman Pasha's small force of 40,000 men at
Plevna threatened the road from Sistova to Kezanlik.
When Napoleon was advancing to Vienna in 1809,
although he had won the battles of Landshut and
Eckmuhl and there was no army ready to stop his way,
he was obliged to weaken his battle-force by large
detachments at Passau and Linz, and in consequence
his army was not sufficiently strong when he came to
the island of Lobau.   Next to coping with the British
armies in the field, the most difficult task of the French
marshals in Spain (1808—1813) was the securing of
their roads to France on the west and east.   Any French
divisional commander could smash up a Spanish force in
a battle; the Spanish generals were absolutely ignorant
of the art of war and too perverse to learn it, or to take
advice as to tactical details, but they knew how to seize
convoys, to close defiles, to break down bridges, to tear
up roads, to hover on the flanks of an army, and to
massacre small parties.   And thus it was that men like
Mina, Longa, and Julian Palarea, better known as the
Medico, rendered it necessary for the French to employ
some 100,000 men between Madrid and San Sebastian,
and Valencia and Gerona.

Some wars are decided at once by great victories.
Sadowa practically disposed of the issues in 1866, and
Marengo of the Austrians in the north-west of Italy in
1800, while Austria yielded after Austerlitz in 1805, but
frequently the defeated troops avoid tactics and resort
to strategy.   France, in spite of perpetual tactical defeat,

and with no regular army, kept the Germans busy from September 2nd, 1870, till February 2nd, 1871—in this case altogether because of a resolute use of population and territory from the strategic point of view. There were no tactics, good or bad, the actions were mere " hard pounding" by the Germans of raw recruits, whose officers did not know ordinary drill, not to speak of tactics.

To sum up the matter:—

The object of a strategist in drawing up his plan is so to arrange his marches and his lines of operations that, on the one hand, if he wins the battle he will not only defeat the enemy on the field but place him in a situation of much perplexity as to his future action, his line of retreat, and his supplies; and, on the other hand, if the battle be lost, he will have secured for himself a safe line of retreat, and an opportunity of recuperating his strength. Better still would it be if he could be so certain of winning the battle that he could venture before the engagement to place himself astride the line of supply of the enemy, who in this case, if beaten, must capitulate. This, however, is more risky, as the situation is reciprocal, and therefore it is generally wiser to compel the enemy to form front to flank—that is to say, to fight with a flank to the base—than to interpose between him and his base. As a general rule it would be found that if *A* manoeuvred so successfully as to interpose between *B* and his base, *ipso facto* *B* would interpose between *A* and his base. When Napoleon in 1800 marched round the right of the Austrians and compelled them to fight with their face to their base, they in turn interposed between him and Mont Cenis and the St Bernard. Had he lost the battle his position would have been one of great danger. When Jackson threatened Banks in 1862 near Strasburg in the valley of the Shenandoah he imperilled the communication of the Federals, but had they stood their ground and beaten him and thrown their right forward he would have been isolated.

It will generally be found in all cases in which the geographical, strategic, and numerical conditions are equal, that success, as in other walks of life, depends on a daring initiative. He who starts first with audacity, and pursues his plan with celerity, secrecy, and resolution, reduces his opponent to the defensive and to irresolution, and shifty expedients have to take the place of manœuvres. The value of time in war is apparent. Opportunity turns the locks on her forehead to the prompt, not to the dilatory. In 1806, when the Prussians were meditating and discussing, Napoleon was marching. While the Federal chiefs on the Chickahominy were wrangling with their War Office, Lee was crossing the river and turning their flanks.

Some generals prefer to divide the operation of interposing between the enemy and his base into two distinct operations, first to threaten his communications, compel him to fight at a disadvantage, and having won a fight, to advance further against his inner flank—that is the flank by which he is linked to his base—and interpose between him and his base; and then, after another victory, either to compel him to surrender or shut him up in a fortress. In the operations round Metz between the 14th and 19th August, 1870, Bazaine's passage of the Moselle was delayed by the fight of Borny on the 14th; both he and the Germans were

across the river on the 15th; on the 16th he was
assaulted on the road between Metz and Mars-la-Tour,
and while wishing to march west was compelled to look
south. Being beaten on the 16th he was obliged to
throw back his right, and the Germans interposed
between him and his object; on the 18th he was again
defeated at Gravelotte and driven inside the forts of
Metz.

Marlborough's plan for 1704 was splendidly au-
dacious, especially as the rival troops were nearly equal
both in numbers and in fighting capacity. He dared to
cast himself away from both his sea and land base, and,
traversing the Rhine and the Suabian Alps, to cross the
Danube, in spite of the resistance of the Bavarian prince
at Donauwörth. Having ravaged Bavaria he recrossed
the river, and found himself face to face with Tallard
and the Bavarians at Blenheim on the Nebel, east of
Dillengen. He thus interposed between the French
and their object, Vienna, but on the other hand he faced
his base, from which, by a projection of the enemy's
left, he would have been completely severed; but, his
tactics being as skilled as his strategy was daring, he
gained one of the fifteen decisive battles of the world,
and, having taken his adversary prisoner, was able to
march back by the Rhine, the Moselle, and Belgium
practically unopposed. Here was a strategic inter-
position, designed months before the battle and carried
out cautiously and secretly, which changed the fate of
Europe—yet all would have been in vain if Tallard and
Marsin had possessed the military genius of Marlborough
and Eugène. The glories of the age of Louis XIV.
were shattered at a blow. Great Britain became a

leading military as well as naval power. Her alliance and the aid of her army were courted by every European potentate from that day till the fall of Napoleon. Such is the power of military genius. It was immediately recognised all over Europe that, in war, brains are better and more valuable than the most heroic courage, though with this quality also the victor of Blenheim was liberally endowed. The force of intellect amid the clash of arms had not hitherto been noted in the current literature of any nation, but the sage poet Addison was quick to perceive it, and the *Campaign* does justice to military art as well as soldierly valour[1].

It is no wonder that the Marquis de Feuquières considered the proper selection of a commander one of the most important and critical responsibilities of a Sovereign or a State. Upon this he writes a chapter— "Of the Care Princes ought to take in Forming Generals for their Service, and how necessary it is for them to gain by their own Experience a competent Knowledge of the persons they design for Command, and to reward them in Proportion to their Services." The value of a sound strategic plan vigorously executed is incalculable: no reward is too great for a successful general ; the only fault for which he cannot be forgiven is failure.

[1] " 'Twas then great Marlborough's mighty soul was proved,
  That, in the shock of charging hosts unmoved,
  Amidst confusion, horror and despair,
  Examined all the doubtful scenes of war,
  In peaceful thought the field of death surveyed,
  To fainting squadrons sent his timely aid,
  Inspired repulsed battalions to engage,
  And taught the doubtful battle how to rage."
  ADDISON, *The Campaign*.

Napoleon, it should be noted, flattered Marlborough by imitating him in the same theatre, directing his march from the Rhine and the Main to Donauwörth, and then turning westward north and south of the river Danube, and taking prisoner the Austrian General, Mack, at Ulm, 1805.

But let us suppose that the manner in which the hostile army is distributed closes all the avenues by which a surprise round a flank or a stroke at the communication could be effected, even then the resources of strategy are not yet exhausted. As the enemy's forces must operate over several lines of advance we can boldly set out to meet him and, before he can gather his scattered fragments together, concentrate against his left and drive it away from the right; then, leaving a detachment to observe the left, turn on the right, and having inflicted on it a decisive defeat, leave another detachment to watch it. Then with the main body we can turn and join the first detachment, and utterly destroy, or drive to a disorderly retreat, the left. The detached body is called "a detaining force," and when the object of the strategist is to break a front with safety to himself, there can be no doubt, to use Hamley's words, that the "skilful use of a detaining force is the principal weapon in the military armoury."

What is a detaining force? A rearguard is a detaining force—that is, its object is to delay the progress or pursuit of the enemy so that the main body which it covers may gain connection with its base, or take a new position, or get home across the frontier; in short, elude the enemy. A good rearguard commander like Davoust, Foy, or Baker is admirable. An advanced guard may be a detaining force if it seizes hold of and keeps occupied the enemy till the main body comes up and engages him seriously.

But to detain (*contenir*) in strategy implies quite a different class of operations. Let us assume that the army of *A* is divided into two sections, and that *B* can defeat either of these alone but would be repulsed by both together. How are they to be kept apart? Manifestly if *B* divided his army into two equal parts his position will not be improved. But suppose he divided it into three parts—two wings occupying a forward position and a central mass upon which either wing can fall back—he may by his left wing keep one half of his opponent's force away from the other for a day or two, during which he can with his right and centre defeat the latter. A weaker force cannot hope to defeat permanently a stronger force, but by the skilful use of natural obstacles, and by deceptive reports and feints, the stronger can be delayed for some time. It is possible for 10,000 men thus to worry (*amuser*, as French strategists put it) 40,000 without any considerable loss for a week. Of course, if the 10,000 could rout the 40,000 so much the better, but this is not its primary duty, and especially it is not its duty to risk defeat; it should be content to manœuvre and to delay. A force employed therefore, not to defeat another force, but to keep the latter so occupied for a short period that it is too busy attending to its own affairs to give any assistance to its comrades in some other part of the theatre of war, say 12 or 50 miles away, is a "detaining force." The left wing may be the detaining force to-day and join the main body to-morrow, the right wing may then play a similar part. But no process except the skilful use of a detaining force could enable an inferior force to escape the disaster that a superior force, if all its parts could be brought together at the same time, could inflict.

Forces operating from divergent bases against their foe, if neither of them be obstructed during their advance, must crush an inferior force if they can assail it in front and flank before the afternoon of the same day. But if either of them be effectively obstructed by a detaining force and prevented joining the other for one day, both may be driven away from each other, and all the labours of a costly and toilsome campaign may be rendered nugatory within sight of victory. That force of which the components act from a common centre outwards, keeping in touch as regards all its parts, every detaining part being able at all times to fall back on the centre, is said *to act on interior lines*. The forces whose parts move from the circumference along the radii of a circle towards the centre in such fashion that if any part be beaten it is driven not towards but away from the other part, is said *to act on exterior lines*. These phrases, though technical, are of such frequent use in all military treatises, especially French and American works, that in spite of the protests of Hamley and others, and the long-drawn substitutes of German writers, it will be found convenient to adopt them. In the campaign of 1812 in the Peninsula, the Allies along the Portuguese frontier were acting on interior lines, Hill being the detaining force. In 1814, Napoleon, though tottering to his fall, furnished all writers on the military art with the most brilliant example of the repeated use of detaining forces on either flank, and the co-operation of the central force at the decisive moment.   His operations and those of his Marshals were altogether magnificent.   Imitating Napoleon, the Confederates practised exactly the same use of detaining forces, with most satisfactory results, in 1862.   A very bad example of what might have been interior lines was Napoleon's operation June 17th, 1815. It would be hard to recall a more instructive instance of the effects of a detaining force than the fashion in which Marshal Davoust wrecked the Archduke Charles' well-conceived campaign between the Iser, the Abens, and the Danube in 1809.

A reader who apprehends the importance of a threat against the enemy's lines of communication, whether by road or river or railway, or by sea; and who understands the vital importance to the defensive, if inferior in strength, of even a temporary segregation of the offensive forces, has sufficiently grasped the Principles of Strategy for the purpose of this treatise.

There may of course, though the public would attach little glory to the operation, be most brilliant strategical successes without any battle.   The strategical combination may be such that the enemy may be obliged to evacuate territory, or, if he remains, be exposed to the risk of capture.

When Graham and Wellington combined north of the Douro in 1813 the French felt bound to abandon Madrid and the Douro without a skirmish, and when our generals changed their base to the Biscayan shore the French also were constrained to abandon Burgos and the line of the Ebro.   In the previous year, when the British entered Madrid, Soult, who was two hundred miles away, found it necessary to abandon Andalusia and the vast accumulation of stores connected with the siege of Cadiz, and to retire to Valencia.

When the strategical plan of operations is settled, the next stage of the proceedings comes within the province of various departments of the General or Head-quarters Staff, all members of which should recognise that they are parts of a common whole, and allow no departmental or class prejudices to arrest for a moment the march to victory. The State must supply each department with ample means; the word economy must be wiped out of the political and military dictionary when once war begins. An attempt to save a thousand pounds to-day may easily result in the expenditure of a million a year hence. Nor does a statesman or a soldier gain any credit for parsimony. If any department fails, as did the medical department of the Americans during and after the late Cuban war, all concerned are over-whelmed with obloquy. Everything, blunders and extravagance included, are forgiven and forgotten forthwith if the soldiers are well treated, and come back victorious and content. The logistic details give far more trouble and anxiety than does the strategic plan, though if this be wrong every effort of the departments will be in vain.

Before placing the army in movement the following are some of the principal points to which attention should be directed. Suitable localities should be selected for magazines, depôts, and hospitals. Plans of the lines of march of each division or corps should be elaborately designed and carefully distributed, so that no clashing or crossing of corps or entanglement of columns should be possible. Sites for camps must be chosen, and villages and towns for cantonments, while the accommodation available therein, and the supply of waggons, draught-horses, and food that each district can afford must be ascertained clearly, or terrible distress to the inhabitants as well as the army will arise. If the inhabitants are not permitted to remain in a state of sufficient comfort, or are half-starved through wholesale requisitions, they will get diseases which will spread among the soldiers. If the soldiers have to sleep in damp quarters or on wet ground, and are under-fed, they will get dysentery. A small saving in food per day may cost thousands of lives. Then if forethought is not devoted to replenishing the supply of ammunition in lavish quantities, battles may be lost. The Turks at Plevna had such an abundance of cartridges that they could fire freely in every direction, and cover with lead all the approaches of the Russians to their fort, and shower bullets over field and wood.

The following details illustrate the enormous requirements of armies in the field. During the bombardment of Algiers 10 British ships fired 39,000 rounds. In all, 500 tons weight of round shot were used, and 966 ten and thirteen-inch shells were fired by the gunboats.

At Plevna 200,000 rounds of shell were fired by the Russians, and 80,000 by the Turks ; the Russian infantry used 10,000,000 cartridges, and the Turkish 15,000,000.

The siege of Strasburg, 1870, lasted 50 days, and the regular attack 31 days, during which time the besiegers fired 193,722 shot and shell on the works and into the town. At Omdurman one British battery is said to have fired 1,000 rounds. A British regiment going into action is provided with 309 cartridges per man, 100 carried on his person, the rest close at hand, or in the ammunition park.

The Artillery material at the disposal of the French army of the East during the Crimean War comprised 1,007 guns, 2,000 gun-carriages, 2,700 waggons, 2,000,000 of projectiles, and 9,000,000 rounds of powder. There were sent to the army 3,000 tons of powder, 70,000,000 infantry cartridges, 270,000 rounds of fixed ammunition, and 8,000 war-rockets.

On the day of the final assault there were 118 batteries, which during the siege had consumed 7,000,000 pounds of powder. They required 1,000,000 sandbags and 50,000 gabions. Of engineer materials 14,000 tons were sent, and the engineers constructed 50 miles of trenches. Of subsistence, fuel, and forage 500,000 tons were sent; of clothing, camp equipage, and harness, 12,000 tons; of hospital stores 6,500 tons; of provision-waggons, ambulances, carts, forges, etc., 8,000 tons: the whole making a total of some 600,000 tons.

It is not necessary to add similar facts for the English, Sardinian, and Turkish armies. Lord Wolseley says :—" The medical history of the Crimean War is a shameful story, and tells of how an army may be destroyed by a ministry through want of ordinary forethought and ignorance of military science : the general can learn from its pages the important lesson that the greater attention he pays to the health of his men, the stronger will be his battalions on the day of battle."

When almost too late the nation awoke to a sense of its responsibility, and the soldiers were at last carefully tended, largely through private charity. Then, of course,

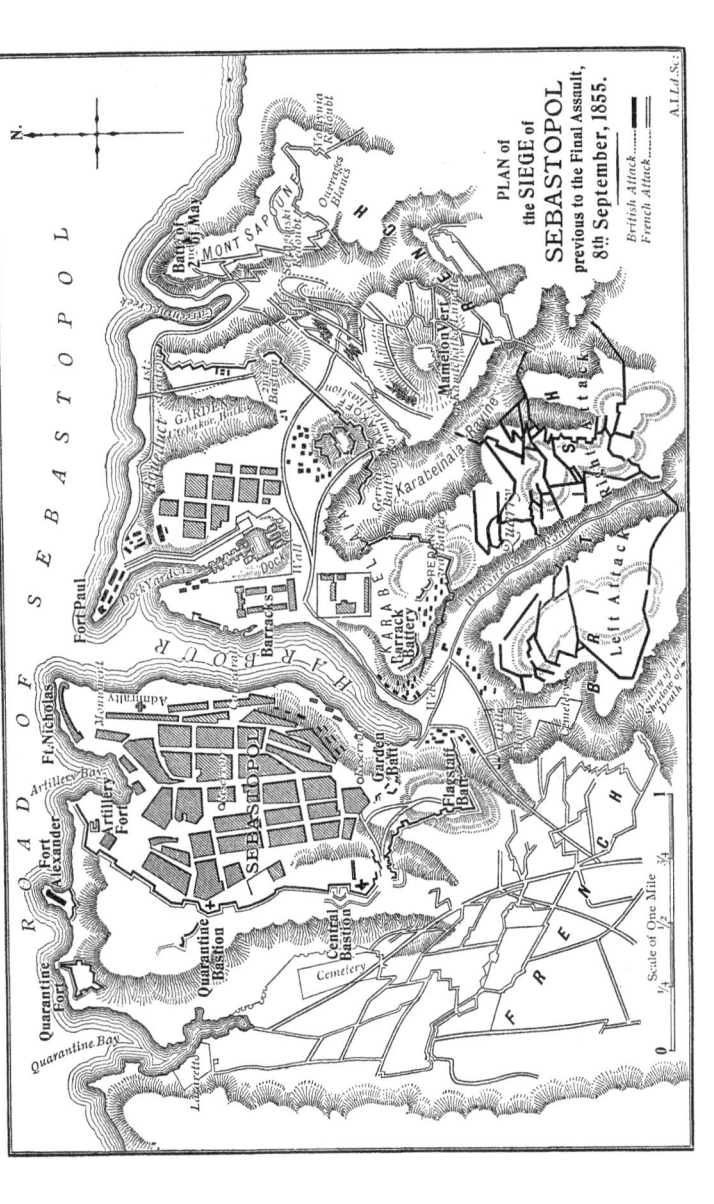

N.

ROAD OF SEBASTOPOL

Quarantine Bay

Quarantine Fort

Fort Alexander

Ft. Nicholas

Artillery Bay

Artillery Fort

SEBASTOPOL

Quarantine Bastion

Central Bastion

Cemetery

Flagstaff Bastion

Garden C. Batt.

Admiralty

HARBOUR

Dock Yard

Fort Paul

GARDEN

Creek Battery

DOCK

Barracks

KARABE Barrack Battery

Mamelon Vert

MONT SAPOUN

Bar of Malakoff

Quarry

Redan

Karabelnaia Ravine

Valley of Death

F R E N C H

B R I T I S H

Left Attack

Right Attack

Scale of One Mile

0   ¼   ½   ¾   1

PLAN of
the SIEGE of
SEBASTOPOL
previous to the Final Assault,
8th September, 1855.

British Attack ....
French Attack ....

A.I.L.J. Sc.

no regard was paid to expense, and every man in the
Allied army cost £200 a year. But reasonable attention to the wants of the troops in time would have been
more useful than all this prodigality, which was too late
to bring back to life 15,000 victims of official neglect.

The enormous requirements of the German army in
France during 1870—71 may be conceived when we
remember that in the course of twenty-four hours each
corps d'armée consumed 18,000 loaves of three pounds
each, 120 cwt. of rice or pearl barley, either seventy
oxen and 20 cwt. of bacon, or a proportionate amount
of prepared sausages, 18 cwt. of salt, 30 cwt. of coffee,
35,000 quarterns of spirits, and 3,500 ounces of orange
essence or some other bitter tincture to mix with the
spirits. To this gigantic repast must be added 60 cwt.
of tobacco, 1,100,000 ordinary cigars, and 50,000 officers'
cigars for each ten days. Finally the forage for the
horses, at a minimum of 20 lbs. for each horse, must also
be reckoned.

Multiply these figures by $17\frac{1}{2}$ and we have the
sum total of the consumption in one day (excepting
as regards tobacco) of the German troops in France.
The difficulties of bringing up such gigantic quantities
of stores were often aggravated by the usual disasters
incidental to warfare. Several times during the campaign, owing to the severe winter, each corps had
distributed among them woollen shirts, flannel bandages,
comforters, plaids, stockings, boots, etc. The field-post
needed a considerable amount of rolling stock. From
the 16th January to the 31st December, 1871, no fewer
than 67,600,000 letters, and 1,536,000 newspapers, in
other words, about 400,000 letters and 99,090 papers

per day were despatched from and to the army.   In the same period 41,000,000 thalers, and 58,000 parcels of all sizes and weights, were sent by the office to the German military authorities in France.   The soldiers received or sent to their friends and relatives at home 13,000,000 thalers and 1,219,533 parcels, or 322,173 of the latter per day.   A large number of sick and wounded were constantly being conveyed back to Germany, besides prisoners, the number of whom was unprecedently large.   Add to all this that towards the close of 1870 from 180,000 to 200,000 new troops were brought up to the seat of war, and that the transport of guns, shell, and every variety of ammunition never ceased for one day until peace was declared, and we can then form some idea of the extreme importance to Germany of having secured command of the various roads and railways.

A noteworthy lesson in logistics was the arrangement of Napoleon's forces during their march from Boulogne to the Danube in 1805, and the orders of Wellington in 1813 were so well conceived that his own divisions and those of Hill and Graham converged on the Zadora in front and to the north of Vitoria at the same hour, to the ruin of King Joseph.   Nor did he neglect any of the other precautions which have been detailed.   He insisted on an ample supply of food, pointing out to the Spanish General O'Donnell that victory was impossible without discipline, and that no troops could be expected to observe discipline who were not regularly paid and fed, and, indeed, he allowed General Hill to use British rations for the support of the Portuguese who were neglected by their own officials.

Such is the complicated machinery for which a General is responsible.

Admirable reports on the state of affairs in the Crimea and on the disorganisation of Gambetta's levées were doubtless to be found pigeon-holed in the United States War Office, and yet, when war broke out with Spain in 1898, nothing was organised except confusion. The arrangements for embarkation at Tampa and disembarkation at Cuba were inefficient to the last degree. The lack of accommodation for the men was not only disgraceful but most injurious to health. Had the enemy been more alert the results would have been disastrous. Lord Roberts says in regard to his Afghan expedition, " Our greatest difficulties on all occasions arose from the want of a properly organised Transport Department, and they will understand that I was able to make this very apparent when the necessity for mobilising rapidly only one Army Corps came to be seriously considered. We were able to demonstrate conclusively the impossibility of putting a force into the field, sufficiently strong to cope with a European enemy, without a considerable increase to the existing number of transport animals, and without some description of light cart strong enough to stand the rough work of a campaign in a country without roads; for it is no exaggeration to say that in the autumn of 1880, when I left Kandahar, it would have been possible to have picked out the road thence to Quetta, and onward to Sibi, a distance of 250 miles, with no other guide than that of the line of dead animals and broken-down carts left behind by the several columns and convoys that had marched into Afghanistan by that route."

As strategy deals with plans of action before the actual collision of armed forces, it is manifestly not affected by changes in weapons or equipments. It is modified, however, by changes in means of communication. The use of steam and the electric telegraph, inasmuch as they facilitate the transport of men and material, and of intelligence and orders, has undoubtedly injuriously affected strategy so far as it depends upon secrecy and surprise. On the other hand, when once the operations have commenced, it has provided generals with powerful agencies for the rapid accomplishment of their aims.

There is much discussion as to the relative merits of an Offensive and a Defensive policy, but there can be no doubt that, if a State be ready and powerful enough, the advantages of the offensive are enormous, and that expeditions should be launched against the enemy when the tension has reached fighting point, whether war be formally declared or not. Suwarrow's motto, *Stupai she*—"Forward and strike"—might be adopted by every military leader.

If a nation is not ready it should not involve itself in political predicaments that may tend to war; it ought to wait, yield somewhat, prepare for a war in peace, and gird its loins for a future struggle. If both sides are quite unready the spectacle is undignified, and much needless torture is inflicted on the soldiers engaged; this was the case in the war between the United States and Spain in 1898, and the feelings and pride of the combatant and non-combatant classes on both sides were unduly strained.

In a naval war the State whose fleets are more

powerful and efficient should forthwith assume the most
vigorous initiative : there are no dangerous defiles on the
main seas, and they should annihilate or drive to port
the weaker fleets at once.   By land also the offensive is
justified by every motive, while the nation and army that
are on the defensive are depressed when they see their
frontier crossed, and strangers forcing themselves into
their villages and seizing the fruits of their labours.
Very frequently in these cases public opinion and the
exigencies of Government compel the staff to adopt
measures which, from a military point of view, are
injudicious.   The French Marshals feared to abandon
the Vosges and retire behind the Moselle at the beginning of the war of 1870.   When only soldiers are
concerned, as was the case in 1815, once the defensive
policy was adopted in Belgium, Wellington and Blucher
were constrained to wait till Napoleon's plan was well
developed before they could make up their minds for
a concentration either on the right, left, or centre.
When, either from lack of resources or preparation, or
geographical difficulties, a nation is bound to remain on
the defensive, it should not be merely a passive offensive,
or a war of positions from which its armies are driven
away step by step ; it should be prepared for a defensive-
offensive, and should be ready to spring upon the foe on
the least appearance of weakness on his part.

In truth it is very difficult to conquer a resolute
State not governed by party shibboleths merely, but
trained in the school of patriotism, loyalty to leaders,
and pride in past traditions.   The frontier once passed,
and the glamour of a few victories over, the bulk of the
invaders cease to be interested, and look back with

craving to their homes.   Their toils increase as they advance into the heart of the country ; they have to wander far in search of food, or be content with the monotonous fare of their own commissariat.   They are embarrassed with regard to their lines of communication, and disheartened at the constant recurrence of illness ; they engage in wearisome sieges, and the marches in long columns cease to be exhilarating.   These become, indeed, the most dreary exercises in which human beings can be engaged, fatiguing to the heads of the column, exhausting to the forces which follow them.   " Slowness and toil," it has been well said, "are the characteristics of the movement of great masses of troops."

As Russia would not yield after the occupation of Moscow, Napoleon was obliged—there being no object to be gained by remaining in a burnt capital—to retreat, and the horrors of the retreat were worse than the terrors of the advance.

In the war against Austria in 1741 Marshal Belleisle entered Bohemia with a fine army, took Prague, and at first carried all before him.   But, before the end of the year, the line of his retreat was traceable by carcases. And yet this very retreat proved that he was not lacking in ability.

In the days of Turenne and Marlborough, armies when overtaken by winter—owing to the bad state of the road, the difficulties of provisioning, and the fact that they were units and not broken up into divisions and corps until after the French Revolution—were compelled to cease campaigning, and disseminate themselves in winter-quarters.   That this was a matter of no small difficulty may be seen from the study of a few campaigns.

In the winter of 1672—3 Marshal de Turenne fixed his winter-quarters in the Westphalian dominions of the Elector of Brandenburg, after he had obliged that prince to repass the Weser.   The precautions he took for their security were elaborate and far-reaching, as we learn from Feuquières.   "The Head-quarters, toward the Weser, were in strong towns, where bodies of horse and foot were likewise posted.   The flat country, which depended on the towns, was divided among the troops who were quartered in those towns, and was appointed to furnish them with their subsistence, as well in kind, as in money; and all the troops of the first line were in the Head-quarters.   Those of the second, who were nearest Lipstat, a town belonging to the Elector of Brandenburg, and in which that prince had a strong garrison, were disposed in the same manner as the first line, with this difference only, that they were obliged to be attentive to the garrison of Lipstat, for their own security.

"Marshal de Turenne, besides these precautions, was careful to mark out a field of battle, at the head of the quarters of the first line, where he fixed the general rendezvous of all the quarters of the army, each of which marked out their routes for their orderly and expeditious arrival at the field of battle, in case the enemy, during the winter season, should attempt to repass the Weser to attack the quarters, which were always secured by these judicious precautions....

"In the winter before the year 1689, and after the conquest of Philipsburg, and the places in the Palatinate, Louis XIV caused one part of his army to winter on this side of the Rhine, and the other part

along the Necker. These quarters were not unmo-
lested, but were raised at the close of January; not
that there was any sufficient reason for that pro-
ceeding, but only through the misconduct of M. de
Montclar, who commanded in the whole extent of that
frontier. The King's troops possessed the Necker from
Tubingen to Mannheim, and consequently the territories between the Necker and the Rhine, except the
city of Stuttgart....

"The imperial garrison of Philipsburg retired, after
that place was taken, to Ulm, and they were all the
troops the Emperor had on this side of Austria and
Bohemia. The circles of Suabia and Franconia had
very few troops in those dominions; the rest were
then in Hungary, where they only began to make the
dispositions for their return to the Empire. In a
word, there were no troops for 60 leagues round the
French quarters who durst attempt to molest them.
And yet M. de Montclar, upon a false intelligence of
the approach of a great body of forces, raised all his
quarters on this side of the Rhine so very suddenly
that it rather resembled the ignominious flight of an
army than the motion of quarters regularly raised[1]."

[1] Feuquières, vol. II. p. 360.

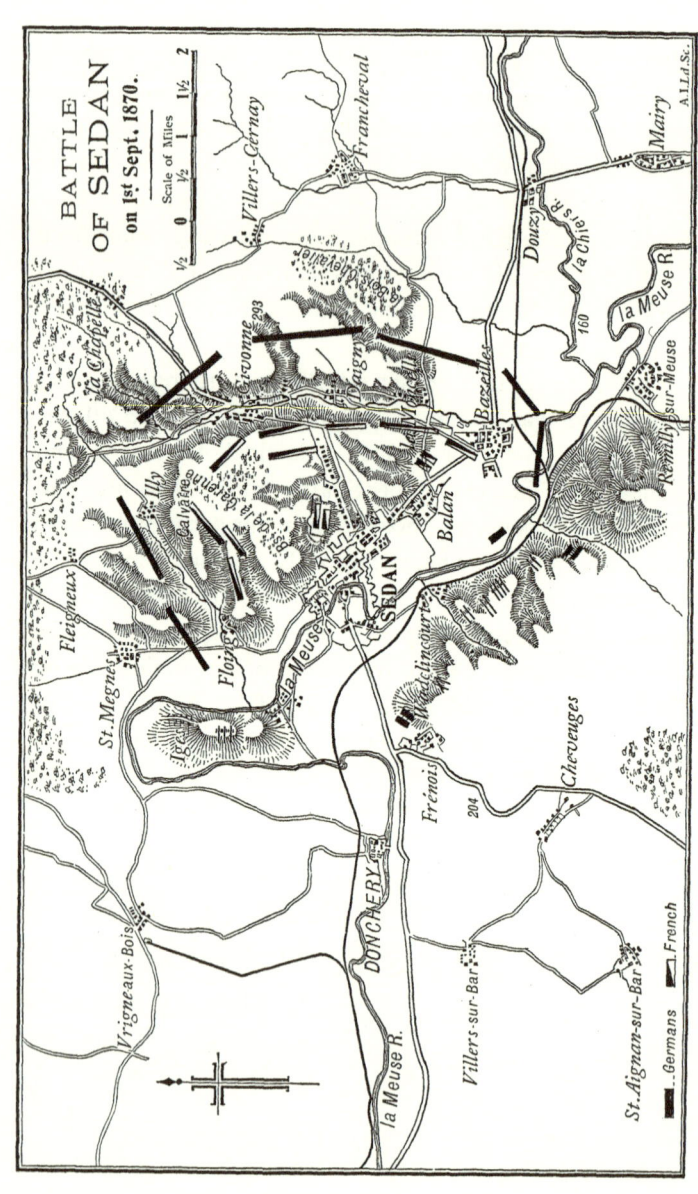

BATTLE
OF SEDAN
on 1st Sept. 1870.

Scale of Miles

½ 0 ½ 1 1½ 2

A.I.I.d.Sc.

Villers-Cernay

Francheval

la Meuse R.

Mairy

la Chiers

Douzy

la Givonne

Daigny

Bazeilles

Givonne

Moncelle

160

la Meuse-sur-Meuse

293

la Chapelle

Remilly-sur-Meuse

la Garenne

Fleigneux

Calvaire

Balan

SEDAN

Bois

Floyng

St. Megnen

Wadincourt

la Meuse

Cheveuges

Frénois

204

DONCHÉRY

Vrigneaux-Bois

la Meuse R.

Villers-sur-Bar

St. Aignan-sur-Bar

▬▬ ..Germans   ◢◣ French

CHAPTER III.

IT is a principle of the Art of War that the general responsible for a given campaign should be allowed to work his scheme out with the aid of his staff to the end, without the interference of anyone—Prince, Cabinet, Parliament, or people. Otherwise success in war is unattainable. When ·it was proposed that the First Consul should direct Moreau's movements on the Rhine and Danube in 1800 as well as his own in Italy, the reply of the only man whose perception of the bearings of Strategy was almost instinctive was, "It is better to have one bad general than two good generals for the same operation," and he was right, as events proved, for Moreau did well, and made a triumphal journey from Schaffhausen to Blenheim, and from Nordlingen to Hohenlinden. If the general be not trustworthy and of assured capacity, then let him not be trusted. Councils of War and councils like the Aulic Council are equally to be deprecated; the former generally wish to shirk responsibility and are pusillanimous, the latter are behind the times and away from the theatre of war. What French critics have called the "*insensée*" flank march from Chalons towards Metz in August, 1870, was not MacMahon's device; it was forced on him by Parisian politicians and the war minister Palikao. He saw that it was a monstrous suggestion, but he obeyed, and as victory was out of his power, he could only face death at the head of his battalions. He was wounded at the dawn of the fatal day of Sedan, when his army, surrounded on all sides, was obliged to surrender.

Mr Pitt senior was our greatest war minister; he preached the doctrine that, "in war, expense is the truest economy." His art consisted in understanding good naval and military schemes, selecting excellent leaders, and then supporting the men of his choice with all the resources of a people whose patriotic enthusiasm his eloquence excited.

It is a significant fact that, during the American War of Secession, for the three years during which the control of the armies of the North remained in the hands of the Cabinet, the balance of success lay with the Confederates. But in March, 1864, Grant was appointed Commander-in-Chief, Lincoln abdicated his military functions in his favour, and the Secretary of War had nothing more to do than to comply with his requisitions. Then, for the first time, the enormous armies of the Union were manœuvred in harmonious combination, and the superior force was exerted to its full effect[1]. The Confederate politicians meddled much less with the arrangements of military chiefs than did the Northern; nevertheless on some occasions President Jefferson Davis interfered, and on one of these Jackson resigned in consequence. Had he not been persuaded to withdraw his resignation the Confederates' strategy would have

[1] Colonel Henderson's *Life of Jackson*, I. 255.

forthwith collapsed.   Had Sherman's advice been taken
at the beginning of the war how different would have been
the result !   When the people at White House said that
a levy of three months' volunteers would crush the
South in one campaign, he replied, " You might as well
put out the flames of a burning house with a squirt."

It is as foolish to begin a war as to build a house
without counting the cost, and the preliminary enquiries
must be recondite and far-reaching.   Before a war begins
there cannot be too much prudence, after it begins there
cannot be too much daring.   In the preparation of any
martial business counsel and care are the principal
elements of success, in the execution of it secrecy and
celerity.   The Chiefs of the State should be thoroughly
well informed concerning the resources of their own
country and those of neighbouring Powers ; the numbers
and military spirit of every nation with which they can
possibly have a *casus belli* as compared with their own ;
the possibility and consequences of combinations of
several hostile States ; the possibilities of alliances, and
how far these may be to the advantage or the detriment
of their own State.   Then the relative finances are to be
considered, the amount of taxation, the pressure per
head having regard to the productive power and income
per head ; the reserve capital of each State, and how
far after paying its own war expenses it could pay a
war indemnity ; how long each can live on its own
resources without importing foreign food or munitions
of war ; how far the leading industries of the nation
would be affected by a war—for example, there was
desperate distress among the mill-hands in Lancashire
in 1861–4 though we were at peace, and bread riots

in Italy in 1898, though Italy was at peace; what then would have been the attitude of the people of each State had it also been at war? These and many other similar questions should be definitely ascertained before the declaration of a war, or the ablest strategist may find himself checkmated by his own people before he has passed the frontier. But for the boundless credit of England how could Wellington have paid his troops in the Peninsula, especially as our Bank had suspended cash payment eleven years before the battle of Vimeira? The credit of France has also done marvels in meeting its own military demands, and in buying the victor out of its boundaries. But could France conduct a long war and pay a heavy penalty for failure with its present phenomenal national debt, the heaviest yet known?

But there are other " sage and serious considerings " that, when once public opinion at fever heat gets control of the national policy, are always ignored. The whole current of the thoughts of the masses runs towards battle. Now it is a proverb among tacticians that, when once a battle begins and considerable sections of the army are engaged, it is impossible to disengage them; the fight, however rashly undertaken, must go on; the other troops must support the first even at the risk of undeserved defeat. So it is with entering upon a war; it must be pursued to the bitter end, however unreasonable or well-nigh suicidal. With a State in such a crisis the only motto is *nulla vestigia retrorsum.* When an association of traders got the final charter of incorporation as the East India Company at the end of the 17th century they little thought that before the close of the 18th they would have become absolute masters of

India. They had no option but to extend their sway from Cape Comorin to the Himalayas, or be themselves obliterated by European or native rivals. Thus, too, the United States, having laid Spain prostrate with regard to the remnants of her colonial empire, while proving their power to conquer a neighbour in one short campaign, have also shattered their own traditional policy and involved themselves in responsibilities of the gravest character beyond the seas. In addition to all these far-reaching potentialities the State that would declare war—or, what is now the same thing, involve itself in a *status belli*—should have the most accurate information with regard to the details of the military condition of its antagonist in every respect, from weapons to organization. This information should be supplied by the naval and military Intelligence Department of the General Staff, if it be skilfully manned and adequately supplied with funds, and, if need be, money should be lavished on procuring by every possible means reliable reports of this description.

The principal lesson of the Franco-German war is that the military institutions of the country,—the army, the materials of war, the fortifications, and the fleet— should be the very first concern of the Government.

The principal military conditions which enter into the case of war are : (1) a field army, ready at all points, and as large as the resources of the country will admit ; and (2) a reserve sufficient to protect lines of communication, to guard fortresses, to complete the numbers of the battalions going to the front, and to fill up the gaps in the ranks which result from active service.

For continental nations conscription of the fullest

character is indispensable; indeed, is the only means whereby sufficiently numerous armies to cope with modern conditions can be maintained. Every healthy man, without any distinction of rank or social position, must be compelled to bear his share of the national burdens. The fact that so many of our own middle and upper classes shirk their military duties in connection with the Militia and Volunteers is a disgrace which may lead to a disaster. The troops must be of excellent quality as well as quantity.

Discipline should be maintained with the utmost severity and inculcated from youth up; in our country drill should be a regular part of the training of every boy, and every adult male should be taught to shoot. The officers should be manifestly of superior tone, of good character, of physical and moral excellence. In time of war punishment should be certain, prompt, and public, in all cases of insubordination. The shorter the service the greater the necessity for absolute discipline.

As to the instruction of soldiers, its object is to prepare men to fight. It ought, therefore, to be adapted to the circumstances of modern warfare. Each individual soldier, every unit of the military organization, ought to be taught to march, to bivouac, and to fight. Military education, then, should be practical, and should, as far as possible, give a clear idea of what would take place in battle.

It should moreover be national, and not a mere imitation of that obtaining in other lands. Every people has its own aptitudes and defects; the former should be developed, the latter cured. Above all, the *morale* of the men should be brought up to the level of their

responsibilities.   The administration of military matters
should be entrusted to thoroughly qualified persons, and
to them only, all other political exigencies being reso-
lutely kept in the background.   Not only should the
military establishments of the nation be complete in
every respect, they should also be instantly available.
The distribution of materials should be easy and rapid ;
the weapons should be of the highest quality, and of the
newest and best pattern.   The fortresses should be in
the most suitable positions, should boast of the latest
devices of engineering, and should be constantly brought
into line with the progress of the military science.
So-called economy is often more foolish than reckless
waste.

It may be well to define some phrases which recur
frequently in all strategic treatises.

The " Theatre of Operation " is all the territory available for the purposes of the war; for example, in 1870 all
France was the theatre of operations.   But manifestly
certain portions of the hostile theatre will suit the purpose of the invader better than other portions ; for
example, it would have been very unwise for the
Germans to advance from between Kehl and Brisach,
while from Sierck to the Lauter was a very judicious
choice.   The term " Zone of Operation " is applied to
that particular portion of the territory selected for the
decisive struggle.   The Germans proposed to limit the
French to the portion of France north of the Strasburg-
Paris road, and this became consequently the zone of
operations.   That their armies had afterwards to operate
in the Vosges, on the Loire, in the Sarthe, and in
Picardy was no part of their original design, and was

due to changes in the political and military conditions of France which they could not have anticipated. In simple language the "zone of operations" is the territory in which, at a given period of time, the general who has the initiative determines to manœuvre. The phrase "theatre of war" is wider than "theatre of operations." All France and all Germany, and the seas on which French or German fleets floated, composed the theatre of war at the end of August, 1870. But in September the theatre of operations became France only, and in December there were no fewer than three zones of operations in France itself. The skilful selection of a particular theatre or zone of operations is a test of strategic ability, and Napoleon's selection of the St Bernard to Milan district of Italy in 1800 was a stroke of genius. There may be two or more distinct theatres of operations in the same war at the same date. In 1814, not to speak of their naval enterprises, the British conducted operations in the north of Italy, in Belgium, and in the south of France, all against Napoleon. In the Russo-Turkish War, one Russian army was at Kars, another at Plevna, and another on the Lom.

The phrase "Lines of Communication" refers to all means, such as rivers, canals, roads, and railways, which maintain the connection between an army and its base, or between the sections of an army which advances by several roads. A "Line of Operation" is the country traversed by the line or lines of communication by which an army advances towards its ultimate object. The base of the Germans in August, 1870, was the Rhine from Coblentz to near Karlsruhe; the line of communication of the several corps comprised all the railways thence to their depôts in the various provinces of Germany; their line of operation the country between Trèves and Strasburg, and thence to the Moselle between Diedenhofen and Toul; their first object the utter defeat of the several French corps that assembled east of the Moselle, and if that would not bring about peace, a vigorous pursuit of the retiring enemy and the investment of Paris. A "Strategic Front" would be indicated by a line linking the heads of the various columns of the army as they faced the enemy. In July, 1813, the strategic front of Marshal Soult was from St Jean Pied de Port to the mouth of the Bidassoa, and Wellington's was from Pampeluna to St Sebastian, and both could only bring together their scattered divisions, or advance against the enemy, by rugged by-ways or dangerous defiles. In 1862 every section of the Federal front looked towards Richmond from Whitehouse to Fredericksburg, thence to Manassas Junction, thence to Winchester and to Franklin.

It is of no small consequence to the strategist whether his adversary's front is, with regard to his own, parallel or oblique, salient or re-entering. In May, 1800, the Austrian front in South Germany was from opposite Strasburg by Freiburg to Stöckach, facing the Rhine from Basle to Strasburg, but Moreau's front was from Schaffhausen by Basle to Strasburg. While he held the Austrians in the Black Forest and won a battle at Stöckach, he so threatened their communications that they retired to Ulm, abandoning 100 miles of territory, and thus illustrating the advantage of a re-entering frontier. Had his front been from Basle to Strasburg, parallel to theirs, he would have had no strategic advantage.

The "Front of Operations" is the space between the strategic front of two armies; this would in modern Europe be very limited in extent, and could be traversed in a few days at most; hence a combat will follow a declaration of war almost at once. The French and German fronts of operation between the Moselle and the Meuse, west of the line Thionville—Metz, and east of the line Verdun—Toul, now almost touch each other.

A front of operations is dangerous from a strategic point of view unless it is developed behind one or more important obstacles; unless it is covered by some artificial defence or natural support on both wings; unless its extent be proportionate to the forces which defend it in line of battle, and these forces can support each other at a critical moment; unless it covers the line of retreat; and lastly, unless, if retreat be necessary, there are behind it within convenient distances good positions for battle.

One of the most interesting branches of Strategic Geography is the fixing of what are known as "decisive strategic points" or localities in the theatre of war, the possession of which by either belligerent will materially affect the chances of success. The student will follow this enquiry all the more eagerly if he realises that, if he fixes upon any locality that *à priori* seems of first-rate strategic importance, in all probability, when he opens ancient history, he will find that his views coincide with those of the Romans. If a line be drawn from the mouth of the Scheldt to the mouth of the Danube as the crow flies it will be found that very little improvement in the knowledge of the relations of geography and strategy has taken place since the days of the Cæsars.

Take Cologne; Cæsar passed from Gaul into Germania
there; Mayence was fortified by Drusus; the French
camp of Châlons sur Marne is the site of one of the
leading battles in connection with the decline and fall
of the Roman Empire. The fortunes of Gaul were
risked at Orléans in the fifth century as well as in the
days of Joan of Arc and D'Aurelle de Paladines. Mur-
viedro was the pivot of Suchet's operations in Valencia
in 1813; under the name of Saguntum it is celebrated
by Livy. The Ticino and the Trebbia are as prominent
in ancient history as in the records of Napoleon. The
course of an invasion of eastern peoples towards Central
Europe and Italy will soon lead them to Belgrade and
the junction of the Save with the Danube, and to Essek
and the junction of the Drave with the Danube. This
is the explanation of the desperate warfare that for
centuries was waged in this theatre between the champions of the Crescent and the Cross, but we find that
the Romans relied on these confluences as splendid
opportunities for keeping the Goths and Huns from the
gates of Italy in the Noric and Julian Alps, and long
lines of fortifications of the most formidable character
may still be traced along their banks. When a reader
for the first time takes up a history of the wars between
the Russians and the Turks south of the Danube he
finds that the country and the names are alike familiar.
Is he not traversing in his mind's eye the zone of the
terrible battles whereby Adrian and other Cæsars for a
period kept the barbarians at some distance from the
treasures of Macedonia and Attica? In 1877 was not
the extreme left of the Russians brought to a stay
before Trajan's Wall, and did not their extreme right

carry Nicopolis ?   It was curious to observe during the
manœuvres in Wiltshire in 1898 that our Commissaries,
without having consulted any antiquarian authorities or
being guided by tradition, planted their principal camps
on the very sites where the Roman legions rested during
their conquest of southern Britain.

The importance of Capitals is not purely accidental ;
it depends upon the very nature of things and often
upon those geological considerations which underlie the
life of creatures.   For example, Paris is really the heart
of France, and was so in the days of Julian and Clovis
as well as of our Henry V. and Napoleon.   Near it are
excellent materials for building, it is on the banks of a
great river, the Seine, and about it converge important
tributary rivers, as the Marne, the Oise, and the Yonne.
It is surrounded by fertile districts, Brie, Beauce, Beau-
vaisis, Le Valois.   All the products of France flow into
Paris, and are thence distributed in all directions.   These
material considerations, rather than the mysterious influences on which Victor Hugo dwells, have made Paris
not only the capital of France, but a centre of civilisation for the world at large.

So, too, what Vienna has been to Eastern Europe
Montreal is to Canada.   It has a population of 140,000,
and covers an area of eight square miles.   The St
Lawrence is here crossed by the celebrated Victoria
Bridge, an iron tubular structure nearly two miles long,
supported on twenty-four piers of solid masonry.   As a
railroad centre, the head of unimpeded ocean traffic, the
outlet of the Canadian system of canals—in brief, as
the connecting-link between the ocean and the lakes
— Montreal is a point of immense commercial and

strategical value, and has been termed the "key and the capital of Canada." In strategic importance it is second to Quebec alone, which is the Belgrade of the great Dominion.

In addition to the principal base there may be *Secondary Bases*.

An army which advances into the heart of hostile territory, as day by day it gets more distant from its base, begins to feel that it is not sufficiently secure. Any further advance might become hazardous, and it is most desirable after a certain number of marches, the precise distance of which will be determined by circumstances, to provide new *points d'appui* before going onwards, as well as a new base.

The defender, on the other hand, as he is obliged to retreat, covers himself with a new line more in the interior. Indeed, France has already a second line with very strong fortresses. In the new works and positions the strategist on the defensive will utilise the rivers and mountain ranges which intersect the invader's line of advance to the capital. If, as is so often the case, the enemy's objective is the capital and it is from 50 to 300 miles distant from the frontier, there will probably be several transverse rivers or ridges, or both, on which stands can be made, but once the invader has passed any considerable river and seized its artificial defences, it becomes a new base for him, and he can stand on it waiting for reinforcements and stores till he finds it convenient to make a fresh advance.

It is all very well for critics to urge rapid marches on generals. No army can move more than 200 miles in 15 days, and very few armies have done as much. Napoleon

thought 210 miles in 20 days a good performance.   For
a few days small armies may manage to cover 15 to
20 miles, but never for a long period, and the reader,
if he wishes to discuss a strategic plan with acumen,
should remember that the average rate of march of
infantry in small bodies is $2\frac{3}{4}$ miles an hour.   A corps
can only march two miles an hour.   Xenophon during
his retreat brought 10,000 men 3465 miles in 215 days
over difficult country.   During their flank march to
Sedan the French only covered $9\frac{1}{2}$ miles a day.

On the 29th of July, 1809, General Crauford joined
the British army which had been victorious at Talavera,
having marched 62 English miles in twenty-six hours.
Each man carried 50 to 60 pounds weight, and there
were only seventeen stragglers.   This was the most
rapid march of any body of foot-soldiers during the
whole revolutionary war.   One of the best French
marches was Clausel's after the battle of Vitoria—
60 miles in forty-six hours.   Cavalry and horse-artillery
can go 5 miles an hour inclusive of halts; field artillery
$3\frac{1}{2}$ miles an hour.

Lord Lake, one of the best cavalry leaders in all
history, though his fame has scarcely reached the ears
of his fellow-countrymen, stretched the mobility of horsemen to the extreme limit before the battle of Ferrucka-
bad.   For some days previous to the battle both Holkar's
men and the British had covered their 25 miles daily
under the burning sun of Hindustan.   But three British
regiments on the night of the 16th November, 1804, surpassed the best performances of Murat, who in 1806
marched 450 miles in 24 days.   They were ordered at
nightfall to surprise the Maharaja's camp.   They reached

the enemy before he could deploy, the horse-artillery opened fire on his tents, and Holkar's army was ruined. He only escaped by an accident. The victors had ridden 73 miles in twenty-four hours by the time the pursuit was over, besides fighting the whole of the Mahratta cavalry.

There were some remarkable marches during the Mutiny, when every Briton in India was on his mettle. On the 13th of May, six hours after receiving their orders, Daly and his men marched out of their station, reached Attock, 30 miles distant, next morning, and on arriving at Rawul-Pindi learned the welcome news that they were to proceed at once to Delhi. On the 9th of June, after moving at the rate of 27 miles a day for three weeks, they marched with a fine springing stride into camp at Delhi, and three hours afterwards went into action with the mutineers. This exploit, it must be remembered, was performed in the hottest part of the hot season. Another instance of extraordinary marching is that of the Sylhet Light Infantry, which, under Major Byng, when in pursuit of the rebels, accomplished 80 miles in thirty-six hours, having started on the 15th December. Finding that the enemy had eluded them, they continued their pursuit after a short rest, and, marching 28 miles more, overtook and defeated him early on the 18th.

Again, in the cold weather of 1858, Brigadier Park, with a flying column, marched 240 miles in nine days, on the last of which he had to thread his way through a thick jungle, and then fought and defeated the enemy. A few weeks later, Colonel Holmes, with a few infantry and artillerymen, marched 64 miles in a little more than twenty-four hours across a sandy desert, surprised the rebels and beat them.   Brigadier Bonner accomplished 145 miles in four days, and Brigadier Somerset 230 miles in nine days.   Such exploits have never been exceeded by any troops; perhaps the nearest approach was a Bulgarian march in 1885.   During the operations near Suakim in 1885 the 2nd battalion of the Grenadier Guards marched 20 miles in a day in rough mountainous country, without a man falling out.

The most rapid continuous march on record of a large army before the introduction of railways was that of Napoleon from the Channel to the Rhine in 1805. Three *corps d'armée* marched on three distinct lines, each corps marching by divisions at a day's interval. The average distance was 400 miles, and the time taken 25 days.   During the Franco-German War the Ninth Corps marched 50 miles in twenty-four hours[1].

Lord Roberts thus describes his arrangements in Afghanistan :—"When it is remembered that the daily supply for over 18,000 men and 11,000 animals had to be drawn from the country after arrival in camp, that food had to be distributed to every individual, that the fuel with which it was cooked had often to be brought from long distances, and that a very limited time was available for the preparation of meals and for rest, it will readily be understood how essential it was that even the stupidest follower should be able to find his place in camp speedily, and that everyone should know exactly what to do and how to set about doing it.

"On the march and in the formation of the camps the same principles were, as far as possible, applied each

---

[1] Hamley, I. Chap. iv.

day.   The rouse sounded at 2.45 a.m., and by 4 o'clock
tents had been struck, baggage loaded up, and everything
was ready for a start.

" As a general rule, the cavalry covered the movement
at a distance of about five miles, two of the four regiments
being in front, with the other two on either flank.   Two
of the infantry brigades came next, each accompanied
by a mountain battery; then followed the field hospitals,
ordnance and engineer parks, the treasure, and the
baggage, massed according to the order in which the
brigades were moving.   The Third Infantry brigade
with its mountain battery and one or two troops of
cavalry formed the rear-guard.

" A halt of ten minutes was made at the end of each
hour, which at eight o'clock was prolonged to twenty
minutes to give time for a hasty breakfast.   Being able
to sleep on the shortest notice, I usually took advantage
of these intervals to get a nap, awaking greatly refreshed
after a few minutes sound sleep.

" On arrival at the resting-place for the night the front
face of the camp was told off to the brigade on rear-guard,
and this became the leading brigade of the column on
the next day's march.   Thus every brigade had its turn
of rear-guard duty, which was very arduous, more particularly after leaving Ghazni ; the troops so employed
seldom reaching the halting ground before six or seven
o'clock in the evening, and sometimes even later.

" One of the most troublesome duties of the rear-guard
was to prevent the followers from lagging behind, for it
was certain death for anyone who strayed from the shelter
of the column ; numbers of Afghans always hovered
about on the look-out for plunder, or in the hope of

5—2

being able to send a Kafir, or an almost equally detested Hindu, to eternal perdition. Towards the end of the march particularly this duty became most irksome, for the wretched followers were so weary and footsore that they hid themselves in ravines, making up their minds to die, and entreating, when discovered and urged to make an effort, to be left where they were. Every baggage-animal that could possibly be spared was used to carry the worn-out followers; but notwithstanding this, and the care taken by officers and men that none should be left behind, twenty of these poor creatures were lost, besides four native soldiers.

" The variation of temperature (at times as much as eighty degrees between day and night) was most trying to the troops, who had to carry the same clothes, whether the weather was at freezing-point at dawn or at 110° F. at mid-day. Scarcity of water, too, was a great trouble to them; while constant sandstorms, and the suffocating dust raised by the column in its progress, added greatly to their discomfort. Daily reports regarding the health of the troops, followers, and transport animals were brought to me each evening, and I made it my business to ascertain how many men had fallen out during the day, and what had been the number of casualties amongst the animals.

" On the 12th August the Head-quarters and main body of the force halted to allow the cavalry and the Second Infantry brigade to push on and get clear over the Zamburak Kotal (8,100 feet high) before the rest of the column attempted its ascent. This *kotal* presented a serious obstacle to our rapid progress, the gradient being in many places one in four, and most difficult for

Passage of the Zamburak Kotal by General Sir F. Roberts.

the baggage animals; but by posting staff-officers at intervals to control the flow of traffic, and by opening out fresh paths to relieve the pressure, we got over it much more quickly than I had expected[1]."

There can be no more striking proof of the strictly scientific character of strategy than its continuity. Napoleon was by his own admission a pupil of Alexander and Hannibal among the ancients, and of Turenne and Marlborough in modern times, and he was eager to be complimented as a hero of Plutarch's school. He wrote an analysis of the campaigns of Alexander, and frequently spoke of the Macedonian's merits during his voyage to Alexandria. The ancient conqueror recognised clearly the importance of a sea base, and the ablest of his adversaries would have probably stopped his career in Asia Minor had it not been for the over-confident obstinacy of the Satraps of Darius. Instead of pushing on at once to the centre of the Persian monarchy after the battle of Granicus, he preferred to form secondary bases on the sea-coast of Asia Minor, for example at Halicarnassus; and he took the maritime towns of Syria and became master of Egypt before risking a campaign in Central Asia. On the other hand, Memnon of Rhodes, the best of the generals of Darius, would have avoided any tactical contest with the Macedonians, being well aware of their superiority on the battle-field. He would have confined himself to strategy; he would have retired slowly before the invader, destroying all provisions as he retreated, somewhat after the fashion of Wellington in Portugal in 1810, and at the same time he would have availed himself of the Persian superiority at sea to make

[1] Lord Roberts, *Forty-one Years in India*, II. p. 347.

diversions in Greece and in Macedonia, just as Bentinck did from Sicily to Catalonia in 1813. But the Satraps concentrated on the Granicus to bar the way to the invader directly, and as a consequence they lost the battle and Asia Minor also, and all chance of utilising their sea power.

The leading points of the works of Vegetius, who lived in the time of the Emperor Valentinian, at the close of the fourth century A.D., might have been quoted word for word out of a 19th century military treatise. His suggestions with regard to recruiting, military exercises, organization, the order of march with advance guards, rear-guards, and flankers, and the skilful use of reserves in battle cause Vial, an excellent French author, to remark that his advice is full of wisdom and philosophic truth, and his treatise, as well as the works of Arrian and Polybius, are far from being mere exercises of archæological curiosity.

Though the movements of the Crusaders were by no means comparable to the great migrations of people which brought about such ruin in antiquity, yet they are of exceeding geographical and strategic interest. As being led by the most expert warriors of Christendom through the territory of the decaying Greek Empire, and as being a bold effort to roll back the tide of Islam from western Asia they are of great educational value, especially when the culture and prowess of the Saracens are remembered. Another important matter in connection with them is that they illustrate the movements of troops by sea, as well as by land, on a large scale. In the middle of the 12th century Louis VII. of France and Conrad II. of Germany were the leaders of a fine

host that went to the succour of the then Christian kingdom of Jerusalem in the 2nd Crusade.   From France
Louis VII. marched to Ratisbon and thence by the
valley of the Danube towards Constantinople, where
he learned the force of the proverb, " Timeo Danaos, et
dona ferentes."   The difficulties of supply were thus
described by the old chronicler :—" From the time when
we entered Bulgaria our spirits and our strength suffered
severe shocks.   Just as we were about to enter a desert
country we procured all the provisions available at
the little town of Brunduse (on the Morava).   Our
stores came for the most part from Hungary, and we
had some difficulty in getting them across the Danube[1].
We suffered much loss in exchanging our coin for
the copper and brass money of the natives.   We had
scarcely carried our crosses into the territory of the
Greeks before these degraded themselves by perjury.
The Greek deputies had sworn unto us in the name of
the Emperor Comnenus that we should find abundant
markets and every facility for exchange.   In place of this
the towns were closed against us, and we could by no
means provide a sufficiency of provisions for the wants
of our folk.   These, then, constrained by penury in the
very midst of prosperity and fertility, were obliged to
satisfy their needs by theft and pillage."   As an example
of how transport by sea was conducted it may be well
to quote an agreement between the Venetians and the
Crusaders, 1201, during the 5th Crusade.   So pleased
was the blind old Doge Dandolo with his guests that

---

[1] On the Danube was an enormous collection of boats brought down by
the Germans, so many that they were used for a long time afterwards by
the dwellers on the banks for building houses and for firewood.

he also took the Cross, and captured, not Jerusalem, but Constantinople. "We shall make *huissiers* [ships with gates on a level with the water-line to embark horses] for the embarkation of 4500 horses and 9000 squires, and other ships for 4500 more horse and 20,000 men-at-arms on foot, and for every horse and every man we shall embark nine months' provision at least. We shall do all this on payment of four marks per horse and two marks per man."

The Venetians kept their agreement so well, even to their loss, that the chronicler wished the barons had made no contract with other maritime towns. "*Ah! quel grand dommage ce fut quand les autres qui allèrent aux autres portes ne vinrent par là! La Chrétienté en eut été bien rehaussée, et la terre des Turcs abaissée.*" The barons on the other hand, at such a distance from their already mortgaged estates, found much difficulty in procuring funds. Although the Count of Flanders and Count Louis and the Marquis of St Paul pawned all their goods, they were 34,000 silver marks in debt, and yet the octogenarian Doge not only forgave them but joined them. All these troubles arose on the way to the Holy Land. It would be most interesting to trace the steps of the armies of warriors who were impelled east-ward by the *volonté de Dieu* from the days of God-frey de Bouillon in the 11th century to those of St Louis at the close of the 13th. How, during the 3rd Crusade against Saladin, were the knights of Richard I. fed, as they journeyed from England and Normandy to Cyprus and Acre, and how did he dare to undertake such an enterprise having regard to the roads and the ships of the period? So full of information are the answers to these and similar enquiries that the Government of France has issued from 1844 to 1880 a magnificent series of volumes in honour of the Crusaders. These volumes also illustrate the ardent desire of the French to be prominent in the East.

As Madame de Witt (Guizot) says[1]:—"A French pilgrim, Peter the Hermit, preached the First Crusade. A Frenchman commanded it, and, during two centuries, the conquest and defence of the Holy Land were intimately connected with the sentiments, the ideas, and the vicissitudes of our country. It was a French king, St Louis, who, last of the crusading chiefs, filled the Orient with recollections of our glory." We shall see in later portions of this treatise how sentimental considerations such as these affect the course of history after the lapse of centuries.

[1] *St Louis et les Croisades*, p. 1.

CHAPTER IV.

COMMAND OF THE SEA.

NO treatise, however elementary, on Strategic Geography would be complete without a chapter on Command of the Sea. Fortunately for our people the labours of so many able and industrious authors have recently been devoted to this subject that the meaning of the phrase is perfectly well understood by every student of modern history, and our politicians of all parties are at last alive to its paramount importance. Shortly, the phrase means the possession of a fleet strong enough and manned by men skilled enough to cope with and defeat an enemy's fleet, wherever it may be found, and when a Power aims at Colonial Empire in modern times, inasmuch as modern war-ships are steam-ships, it means not only frequent 'opportunities of port' for provisions, but also secure coaling-stations on every ocean route.

States have gained command of the sea by defeating hostile fleets, and have then proceeded, both in ancient and modern times, to found colonies and to land military expeditions either to harass or to conquer foreign communities. The science of this branch of strategy is clearly set forth in a most convenient form in the works of Sir George Clarke and Messrs Thursfield and Wilkinson, not to speak of Admiral Colomb and

Capt. Mahan.   The present chapter is concerned rather
with that which actually took place, than with what
ought to have taken place had the leaders on both sides
made the most of their opportunities, and had foresight,
combined with experience and incorruptible patriotism,
prevailed among the Governors of the various races of
mankind.

In modern times, in spite of railways, good and
numerous roads, and canals, the sea remains the
main agency for the spread of civilisation and the
circulation of commodities.   In ancient times land
journeys were everywhere beset with dangers and
encumbered with obstacles, and, till the Roman roads
connected together the cities along the Mediterranean
and cut into the heart of Thrace, South Germany, and
Gaul, sea routes, though confined to inland seas and
narrow waters, were still of the greatest importance.

All naval annals begin with the Mediterranean.
First among the races that claimed command of the sea
were the Phœnicians.   They founded their settlements
on the coast of Palestine at a time when the Egyptians
were at the height of their power, having previously
been daring navigators of the Red Sea, and perchance
of the Persian Gulf.   The forests of Lebanon furnished
the sturdy merchants of Tyre and Sidon with ample
material for their ships.   The scientific observations of
the Chaldean star-gazers were utilized to guide them
over the deep, and their kinsmen, the Canaanites, were
doubtless ready to add to the numbers of their soldiers
and colonists.   Their energy, daring, and success from
the time of Homer and the Mosaic writers till the days
of Alexander the Great are commemorated by all ancient

historical writers, sacred or profane ; they facilitated communications between the Western and Eastern Mediterranean and their stations were the *entrepôts* for the luxuries of Asia. Spain and north-western Africa proved sources of wealth ; they sold amber from the shores of the Baltic and tin from the mines of Cornwall. They had colonies in Malta, Sicily, Marseilles, by the banks of the Guadalquivir and in Cadiz ; they founded Carthage and, under Necho, are believed by many authorities to have circumnavigated Africa. The mercantile city was able to withstand Nebuchadnezzar after his conquest of Jerusalem for thirteen years. It resisted Alexander the Great for seven months, in spite of his enormous mole driven to the island through the sea, but when the Persians dismissed their fleet he gained sea power, and the independent political existence of Phœnicia was at an end 332 B.C. The diversion of trade to his new city Alexandria accelerated its decay. With the exception of Beyrout the harbours on the coast are now silted up, and few traces remain of the ' city of purple '—' the city that was glorious in the midst of the sea '—'the crowning city, whose merchants were princes, whose traffickers were the honourable of the earth.'

But while the influence of Tyre in the eastern waters of the Mediterranean was waning, its offshoot in the west was rising in opulence and power. It is a singular fact that the Egyptians, clinging to the valley of their sacred and life-giving river, never thought of sea power, indeed regarded mariners as ' accursed beings.' While their caravans traversed the less hospitable deserts and the Nile was covered with craft they had not a ship on the sea.

But the Phœnician flag was triumphant many a

time for many a generation after the exploit of Alexander.   Indeed the heroic period of the Carthaginians was contemporary with the rule of his generals and their descendants, from Seleucus to Perseus. Having mastered north Africa from the Syrtis to Ceuta, they dared to pass the Pillars of Hercules and to voyage south as far as the mouth of the Senegal river and north as far as Britain.   They commanded the Sicilian strait by the possession of Malta, and it was when endeavouring to become masters of Sicily that they came into collision with Pyrrhus and Timoleon and their fatal foe the Romans.   The fertility of Sicily was celebrated, and accordingly the Carthaginian merchants looked upon it with longing eyes, and when Rome grew in population she derived the greatest part of her corn and other necessaries from this most productive island.   The Romans, hitherto merely land warriors, saw the value of sea power, and Duilius Nepos equipped a fleet of twenty-five galleys and defeated Hamilcar Barca in B.C. 260,—the first naval victory ever gained by them. At the close of the First Punic War the Romans were a leading sea power.

Hannibal during the Second Punic War made the Carthaginian name for ever glorious among strategists and tacticians, but his genius was displayed on land ; and the Romans alike in Sicily and in Spain carried all before them, and having gained command of the sea, under the leadership of Scipio Africanus attacked Carthage itself.   This master-stroke recalled Hannibal from Italy.   After he lost the battle of Zama 202 B.C. one of the conditions of peace was that the fleet should be destroyed.   When the Romans captured Carthage finally in 146 B.C. they also captured Corinth, and thus had undisputed sway of the sea.

The predominant position so long held by the Greeks, and all the glory of Athens, depended upon sea power. The battle of Salamis 480 B.C., which was contemporaneous with a great defeat of the Carthaginians at Himera in Sicily, rendered Asiatic rule impossible in Europe till the arrival of the Turks, and by giving the Grecians the command of the Aegean, opened up a way for such expeditions as those of Xenophon and Alexander.

The position of Athens from the date of Salamis till the Peloponnesian War B.C. 431 was one of the proudest preeminence, and its history at this epoch proves—as does also Roman, French, and British history—that success in war, whether by land or sea or by both, conduces to mercantile, literary, and artistic success; while a nation that forgets courage, 'the chiefest virtue,' and loses skill in the arts of war soon lags behind in the arts of peace.

Athens resembled England in being an excellent founder of colonies, and its colonies resembled some of ours in being ready not only to revolt from, but to wage war against, the mother country. The policy of Themistocles not only saved Athens; it gave to Greece her place in history. After the defeat at Salamis, Xerxes, fearing for the safety of his communications, hastily retreated to Asia, whose coasts were at the mercy of Greeks for generations. The Confederacy of Delos was designed to give the Greeks permanent naval power over the waters of the Aegean, but this confederacy was used by Athens for selfish purposes. Pericles continued the policy of naval supremacy. But at the close of the Peloponnesian War the defeat of Conon by Lysander at Aegospotamos 405 B.C. gave the Spartans supremacy for a short period, till they were in turn defeated by Conon at Cnidos, 394 B.C. He then restored the fortifications of the Piræus, but too late ; the prestige and power of Athens had departed.

The Romans made the Mediterranean an Italian lake, and had no competitors by land or sea from the days of Scipio Africanus the younger to the irruptions of the Barbarians. But they were not a maritime race by instinct, and fell far below the commercial enterprise and spirit of adventure that distinguished the sailors of Tyre and Carthage, or the followers of Nearchus, whose geographical discoveries did such credit to his employer Alexander.

In the civil wars which so frequently broke out from the days of the First Triumvirate to those of Constantine the command of the sea had a decisive influence. At the close of his career Pompey's power depended upon naval superiority, and at last, when beaten at Pharsalia, he was able to escape to Egypt.

In the *Epistles to Atticus* Cicero, speaking of Pompey's military plans, says :—"Cujus nunc consilium Themistocleum est, existimat enim, qui mare teneat, eum necesse esse rerum potiri,"..."navalis apparatus ei semper antiquissima cura fuit."

The Romans made few improvements during their long supremacy either in ships or in the art of navigation.

During the records of the Eastern or Byzantine Empire, for a thousand years, we find the sea power of which Constantinople became the centre was the safety of the State. As long as they held control of the Dardanelles and the Bosporus, and preserved the secret of the Greek fire, not only were the degenerate Greeks in the Sea of Marmora able to devote themselves to games and luxury with impunity, but neither Saracens nor Turks could secure a permanent footing in the Mediterranean Isles.

In the dark ages the Visigoths gained command of the sea and the hardy Scandinavians harassed the coasts from Denmark to Italy. In the middle ages enormous wealth was poured into the coffers of the Italians by the crusaders, who required transport to the Holy Land. The cities of Italy—Venice, Genoa, and Pisa—became rich and ambitious, supported large navies, and contended with the Saracens and Turks and with each other for predominance in the Mediterranean. Some of the enterprises of the Pisans, such as capturing with their fleet of three hundred vessels the Balearic Islands from the Saracens, and their attack on Palermo, are justly celebrated. Genoa still honours the memory of the great Doria, and the " Blind Old Dandolo " planted the standard of Venice on the ramparts of Constantinople.

The Venetians fought against western as well as eastern Christians. After the Venetian fleet under Ziani had captured or destroyed the fleet of Otho, son of the Emperor Frederick Barbarossa, Pope Alexander III. presented the Doge with a ring, using these words:— " Use this ring as a chain to retain the sea henceforth in subjection to the Venetian State; espouse her with this ring, and let this marriage be solemnised annually, by you and your successors, to the end of time, that the latest posterity may know that Venice has acquired the empire of the waves and holds the sea in subjection in the same manner as a wife is held by her husband." This annual marriage continued till the independence of Venice was destroyed, and any of its outlying possessions that had escaped the Turks were seized by Bonaparte.

Marco Polo was a Venetian noble ; he spent some years in the cities of Asia Minor, and then visited Hindustan, China, Japan, Ceylon and other islands, and amazed the Europeans by his vivid descriptions of vast territories, mighty cities, and stores of wealth of which they had hitherto never dreamed, but which they now determined to seize. In 1420 the Venetians had three thousand trading-vessels, three hundred larger vessels manned by 8000 sailors, and forty-five large galeasses, while no less than 16,000 carpenters were employed in their arsenals. After the fall of Constantinople the Venetian power was at its height, but it began to decline after the voyages of Columbus and Vasco da Gama.

The armaments which besieged medieval ports would have seemed very paltry affairs to naval officers like Rodney or Nelson, Hornby or Seymour. Edward III. was able to command the coasts of northern France and the Bay of Biscay with vessels which carried on an average some twenty mariners ; the Mayor and Sheriffs of London, and the authorities of every seaport town were compelled to provide vessels of forty tons and upwards, and to furnish them with armed men and all warlike necessaries. Edward had some five hundred of these, and his sailors dared to attack and board much larger and more powerful vessels from Spain and Italy, and the Hanseatic Towns. For the blockade of Calais he assembled no fewer than seven hundred and thirty-eight English, and about forty other ships carrying some 15,500 mariners. He began the ' Hundred Years' War ' with a great maritime success off Sluys. The Turkish occupation of the Levant closed the old commercial routes to the East, or at least rendered them most unpleasant for Christians, and this gave a stimulus to enterprise in other directions. Towards the end of the 15th century—when France was settling down after the long-continued inroads of the English, and England was reposing after the Wars of the Roses—the marvellous energy of Portuguese mariners led to the discovery of the Cape of Good Hope in 1487 by Diaz, and the attaining of India by Vasco da Gama in 1497. In the beginning of the next century were the splendid records of Albuquerque in India, and expeditions to such distant places as the Rio de la Plata, Goa, the Moluccas, Ascension, and Ceylon.

The Spanish having expelled the Moors, and being united under Ferdinand and Isabella, were also ready for a share in the wealth and excitement which daring navigators were opening up to western Europeans. Columbus steering for the East Indies came across the exhaustless riches of the Caribbean Islands, and his successors Cortes and Pizarro presented to Spain the wealth of Mexico and Peru. In 1493 Pope Alexander VI. acknowledged the sea power of both Portugal and Spain by dividing between them all parts of the world yet undiscovered. The Pope's original delimitation was that all territory west of an imaginary line drawn 100 leagues to the west of the Azores and Cape Verde Islands should belong to Spain, and all east thereof to

Portugal, and by the treaty of Tordesillas in the following year the line was removed to 370 leagues west of
the Cape Verde Islands. The growth of the empires of
both was rapid and far-reaching, and it was entirely due
to sea power. Spain and Portugal were united from
1580 to 1640, and their empire began to fall to pieces—
a prey to the northern nations of Europe, the English
under Elizabeth, and her allies the revolted Dutch subjects of Spain. The French also soon began to develope
commercial and colonizing activity at the expense of
the Iberian Peninsula. The vast empire of Philip II.
and his successors in the Mediterranean, Atlantic, Indian,
and Pacific Oceans could only be maintained by command of the sea; this was lost in 1588, and, as Lord
Bacon says, the Spanish Empire has ever since been
similar to the Roman Empire during its decline and
fall, " every bird taking a feather." No land armies
could save it, nor fortifications, nor wealth. Dissatisfied
colonies discarded the feeble old country, and more
powerful fleets chased its flag from the open waters,
insulted its home ports, and took over its trade, while
pirates like the Buccaneers or admirals like Anson seized
Spanish galleons full of the precious metals.

The fall of the Portuguese empire supports the
theory that " trade follows the flag." The Dutch, after
beating the Portuguese in India, took over their lucrative
Indian trade. They did this the more easily as they had
a better mercantile system ; they not only brought goods
from the East to their depôts in Holland, but they
established distributing depôts in every other country in
Europe, and put the commodities at the doors of their
customers, whereas the Portuguese brought goods to

Lisbon, whence the customers had to convey them to their own depôts. Another blow to the Portuguese was the closing of Lisbon to Dutch merchants by Philip II.

From the days of Cromwell's wars with the Dutch till now the command of the seas has been a branch of British history. The glory of the British Navy may be followed in one continuous track of light since his time. He compelled the Spaniards, Portuguese, and Dutch to recognise the principle that territorial claims were not valid where there was not effective occupation; he put an end to the Portuguese monopoly in 1654, repudiating the old claims founded on the Bull of Pope Alexander VI. He opened the Eastern seas to British trade, and exacted reparation from the Dutch for their massacre and torture of British adventurers in the Spice Islands.

It is clear then from history that not only is an insular Power secure as long as it has command of the sea, but that it can assume the offensive with military expeditions to every part of the world, and thus take "as much or as little of the wars as it pleases," either being content with passive defence for a time, or harassing the coasts and roads, and hampering the commerce of its rivals. But a merely defensive policy, if long maintained, has always been disastrous. A naval Power that cannot keep up a large navy to strike prompt and hard blows against any aggressor in any sea that borders its domains will soon perish. In 1692 the battle of La Hogue saved our empire; in 1759 the battles off Quiberon and Lagos, and in 1782 the battles off St Lucia and in the Indian seas. The ultimate fall of Carthage was due to its lack of a sufficient navy; when this was deficient not even the genius of Hannibal could give security.

Admiral Colomb[1] proves conclusively that the use of steam and other mechanical appliances in the Chino-Japanese War in 1894, and the Spanish-American War in 1898 has in no wise affected the strategy of naval warfare, any more than rifles and breechloaders have affected strategy on land. Whether the Admiral be Doria or Barbarossa, Hawke or Sampson, Graves or Cervera, whether competent or incompetent, his success or failure will in each age depend on the adoption or neglect of a few leading principles. These the American officers, who manifestly had mastered the schemes of operations which made Nelson the embodiment of sea power, applied with skill and vigour in the late war, and thus they won, while the Spanish fleets drifted, rather than manœuvred to their doom.

The principal functions of a fleet have always been to blockade hostile squadrons in their own ports; to have a reserve squadron close to their own shores in order to prevent panic and to protect their own harbours; to convey and safeguard the mercantile marine ; to protect 'provision ports' to which are now added coaling stations; but above all things to be able to meet and destroy battleships of the enemy in a great sea-fight.

In connection with naval strategy much has recently been said about *guerres de course*, an operation of war likely to be renewed in consequence of our foolish adherence to the Declaration of Paris in 1856 in a temporary fit of adulation for France, which now proposes to revive against us its privateering policy of the last century. The proximity of France to England facilitated commerce-destroying operations. Having ports in the

[1] *Journal of the United Service Institute*, April, 1899.

North Sea, in the Channel, and on the Atlantic, her cruisers started from points near the focus of English trade both coming and going. Dissemination of ports was an advantage rather than a drawback to privateers who did not propose methodical and combined regular sea warfare, but rather individual and fitful efforts. As the United Kingdom now largely depends upon external sources of food-supply, it follows that France is the nation most favourably situated to injure it by harassing its commerce. The position is stronger than it was in the last century, Cherbourg presenting a good Channel port, which France lacked in the old wars. On the other hand steam and railroads have made the ports on the northern coasts of the United Kingdom more available, and British shipping need not, as hitherto, focus about the Channel.

The following is a summary of the naval strength of England, France, and Russia in March, 1899.

	England	France	Russia	France and Russia	English strength necessary to blockade France alone
Standard Battleships	63	40	28	68	66
Armoured Cruisers... (modern)	21	16	5	21	26
Older Ironclads and Coast Defence......	24	15	15	30	---
Modern Cruisers......	116	40	18	58	90
Torpedo Craft.........	326	300	236	536	—

In the above comparison France and Russia are selected as the two next strongest Powers after ourselves. The lowest standard laid down for England by various experts is one of equality to the two next strongest Powers. This table will show our position with regard to that standard. In cruisers we are above it; in battleships below.

Of the 63 English "standard" battleships, 10 carry as their main armament muzzle-loading guns, and one is partly armed with muzzle-loaders. These guns are old and take different ammunition from the breechloader. They use black powder and not cordite—the disadvantageous consequences of which were felt keenly by Americans in the war of 1898. They are necessarily weaker than modern breechloaders, and are a serious cause of complication and inconvenience. Since modern heavy guns are replacing the old patterns in the French fleet, it is no longer true that our muzzle-loaders will have to face old weapons in other fleets. All except one of our old ironclads are armed with the muzzle-loader, and in many cases carry no larger quick-firer than the 6-pounder; that is to say they are without the 6-inch or 4·7-inch guns which the Japanese found of such immense value at Yalu.

Our coaling stations, starting from the St Lawrence and coming round to Vancouver's Island are:—Quebec, Halifax, Bermuda, Barbados, Jamaica, St Vincent, Trinidad, St Lucia (omitting home ports), Gibraltar, Malta, Aden, Karachi, Bombay, Trincomalee, Seychelles, Chagos Islands, Mauritius, Diego Garcia, Singapore, Hong-Kong, Wei-hai-wei, Labuan, Thursday Island, Brisbane, Sydney, Adelaide, Melbourne, Hobart, Albany,

Auckland, Wellington, Christchurch, Dunedin, Fiji, Durban, Simonstown, Cape Town, St Helena, Ascension, Sierra Leone, Falkland Islands, Esquimault.

The time taken for the voyages of the best appointed modern mail steamers from London to Bombay is about 13 days, to Hong-Kong 24 days, to Shanghai 28, to Adelaide 26, to Cape Town 17. A fast passage between London and New York takes 5 days and 8½ hours. It must be remembered that all the early navigators sailed in small ships, wretchedly equipped, and that our Indian Empire was founded when the voyage out took several months. The rate of 350 miles a day is a good average now for a steam-ship.

Vasco da Gama's journey to Calicut and back took two years, two months, and five days, during which some time was occupied in collecting samples of Indian products. Magellan, the first circumnavigator, was three months and eight days crossing the Pacific from his Strait to the Ladrones, and the whole voyage occupied just under three years. Drake sailed from England on the 17th December, 1577, with 166 men and five small vessels, and returned November 3rd, 1580, after his wonderful circumnavigation, with one ship and about fifty men. On one voyage he made a run from Florida to the Scilly Isles in 23 days. Anson's celebrated voyage round the world lasted about four years; he brought back in the *Centurion* the ruins of his force, and an enormous booty. Warren Hastings' first voyage to India lasted from January, 1750, to October in the same year. Nelson's trip to the West Indies and back began May 11th, 1805, from Lagos Bay; he reached Antigua June 12th, left it the following day, and was at the

Azores July 8th, returning to Gibraltar July 19th. Lord Roberts started for India February 20th, 1852, and taking an overland route through Egypt reached Calcutta April 1st.

With regard to cheapness of carriage, the cost of a given weight over a given distance by road is ten times the cost of the same weight for the given distance by railway, and the cost by railway ten times that of the same weight for the same distance by sea. Thus, 50 tons by road costs as much as 500 by rail, and 5,000 by sea.

Time is a supreme factor in war, and a nation which is not ready when hostilities break out, may never be able to get ready, but will be exposed to a succession of defeats, however great its wealth, numbers, and capabilities. Therefore countries which object to a large military expenditure and huge establishments in time of peace should aim at an organisation which would give time to develope their military power and change a passive defence into an offensive policy, otherwise, with all the potentialities of success at their disposal, they may be obliged to submit, not only to humiliating conditions, but to such guarantees and charges and loss of decisive posts as will cripple their energies for a considerable period. So much for organisation ; once the campaign begins the mottoes of Gustavus Adolphus and Suwarrow, " Time is the first thing," and " Forward and strike," based, as they are, on experience, are the best upon which to act.

A case of the folly of entering on a naval war unprepared, almost as instructive in its way as the Franco-German War, was the war about impressment and the right of search between the United States and

England in 1812–1814. In spite of a few much belauded
victories by American frigates over British ships, the
difference between a successful and an unsuccessful
maritime war could not be more strikingly illustrated.
In Alison's words — "Perhaps no nation has ever
suffered so severely as the Americans did in that
war. The foreign trade anterior to the estrangement
from England (*i.e.*, 1812)—£22,000,000 exports and
£28,000,000 of imports—was literally speaking annihilated, for in 1814 the exports had fallen to 1,400,000,
and imports to less than £3,000,000. Two-thirds of the
mercantile and trading classes were insolvent, while our
exports and imports, which in 1812 were £64,000,000,
had increased in 1814 to £87,000,000."

Notwithstanding this increase of about 20 per cent.
in our trade, a recent American author states that our
trade was ruined. He may be answered by another
quotation, but this time from an American historian,
Patton, who thus writes of the result of the war :—
"Affairs were most desperate, the treasury exhausted,
the national credit gone, the terrible law of conscription
like an ominous cloud hanging over our people ; civil
discord ready to spring up between the States, coasts
yet subject to marauding expeditions, while the inhabitants were crying vainly for relief." The Legislature of
Massachusetts, "after recapitulating the evils which war
had brought on the people they represent, expressed
sentiments on other wrongs such as enlistment of minors
and apprentices, the national government assuming
command of the States Militia, especially the proposed
system of conscription for both Army and Navy."
" Strange proposition," writes Admiral Sir V. Hamilton,

"for a government professedly waging war to protect
its subjects from impressment[1]."

Dakar and Walvisch Bay on the west coast of Africa,
and Obok and Delagoa Bay on the east coast are
strategic positions of much importance with regard to
our Imperial communications. Though St Louis is the
capital of the French settlement of Senegal and the
most important place between Rabat in Marocco and
Sierra Leone, its trade, owing to its bad anchorage,
is carried on by Dakar, which is connected by rail with
the capital, and has a deep harbour completely sheltered
from westerly gales by Cape Verde, and defended by
the fortified and historic isle of Goree. With regard to
the German ports on the west coast of Africa it is clear
from Prof. Keane's account[2] that neither Sandwich
Haven nor Angra Pequena Bay can give us much
trouble, and fortunately the British have possession of
Walvisch Bay, with an area of 700 square miles. The
British and Cape Governments are determined that this
vitally important position must be kept at all cost,
otherwise a region of 400,000 square miles would be
lost, and it would in foreign hands be a base for operations against all our territory between the Zambesi and
Orange rivers.

Obok has great advantages as a port for steam-ves-
sels. Situated, as it is, in close proximity to the Straits
of Bab-el-Mandeb, it commands this ocean passage within
a much closer distance than does Aden. Merchant
vessels, too, can coal at Obok without having to alter
their course for this purpose, as they may have to do, if

[1] *Nineteenth Century*, March, 1896.
[2] Stanford's Compendium—*Africa*, vol. ii.

they go to Aden. It is true that as a port Obok cannot compare with Aden; still it has a good anchorage, and without entering upon extensive works, it could be converted into a perfectly safe roadstead, for it is sheltered from the open sea by a line of coral reefs, through which canals or passages could be opened out and made practicable for large vessels. North and north-east winds—the most dangerous for navigators in this part of the world—are deflected from Obok by Capes Kolodtsa and Ras-el-Bir, which project far out seawards to the north of the port.

Delagoa Bay is the finest natural harbour and one of the most important strategic positions in southern Africa. The chief town on its shores is Lourenço Marquez. From Lourenço Marquez a railway runs by way of Pretoria to Johannesburg, a distance of about 400 miles; but the nearest point of the Transvaal is only 50 miles away. As a glance at the map will show, Delagoa Bay lies at the mouth of the Mozambique Channel, through which our ships must pass going and coming between the Cape and our possessions further up the east coast of Africa, so that if this port fell into hostile hands our shipping in East African waters would be in grave peril. How grave a peril, is shown by our losses in the great wars of Nelson's time when the French held Mauritius. Between 1795 and 1797, though our fleets were everywhere victorious, we lost between the Cape and India, to French cruisers, no less than 1,475 merchant vessels; and we went on losing till 1810, when we captured Mauritius, and our losses stopped altogether. The injury inflicted on us a hundred years ago from Mauritius might easily be repeated to-day from Delagoa Bay.

Even now, in time of peace, it is used to cripple our trade, and to give foreign goods an advantage over ours.

Under a treaty made in 1823 with the native kings, half of Delagoa Bay once belonged to England, but Portugal claimed the whole of it. Eventually the dispute was submitted to the arbitration of the President of the French Republic, who gave his award against us. But on the 17th June, 1878, a week before the award was declared, the Portuguese Government bound themselves " not to cede or sell to any third Power the territory known as Delagoa Bay, without giving Her Britannic Majesty's Government the opportunity of making a reasonable offer for the purchase, or acquisition by other arrangements satisfactory to Portugal, of the territory thus awarded." In 1891, under a Treaty negociated by Lord Salisbury, this pledge was repeated and extended to the whole of the Portuguese territories south of the Zambesi. Thus it is that England enjoys rights in Delagoa Bay such as no other Power can lawfully claim or set aside.

But there are two Powers, Germany and the Transvaal, who by their words and actions indicate something more than a reluctance to recognise the finality of our right to the pre-emption of Delagoa Bay. The Transvaal naturally wants the harbour, having no outlet to the sea. And we shall do well to keep in mind the fact, for fact it is, that not long ago Germany is reported to have tried to land troops at Delagoa Bay for service in the Transvaal. That Portugal will loyally abide by the Treaty of 1891 need not be doubted, and if the naval power of that country were comparable to that of the greater Powers our rights might be held secure. But the navy of Portugal is the reverse of powerful.   Indeed if the British fleet were not behind her, her rights in Delagoa Bay would be insecure, and her loss of that port but a matter of time.

From the point of view of sea power the province of British Columbia is of great value to the Empire at large as well as to Canada.   Save for its many fine ports we have no naval position from Cape Horn to Bering Strait. It seems strange that this immense stretch of coast for 115 degrees of latitude should have been forgotten, and that we should have omitted to seize some port on it, especially as the importance of the Pacific had been foreseen by Drake, but it is stranger still that British politicians in the sixties were quite willing that British Columbia should have been given up to the United States.   Even as Nova Scotia stretches out towards Europe, so does this province stretch out towards Asia, and the distance between Liverpool and the far East by way of Vancouver is 1000 miles shorter than by way of New York and San Francisco.   Moreover the ocean current, which leaves the coast of Asia, flowing eastward to the American continent, gives to ships bound for a north-western port a gain of twenty miles every twenty-four hours[1].

That the British methods as to colonies after they obtained sea power have been far more excellent than any other methods known is proved by history, and is admitted in a very able book on *The French in Further India*, by M. Chailleul Bert.   Of course India and Egypt can scarcely be called colonies; they are territories under our rule, admirably administered.   But with regard to

[1] "The Greatness of Canada," *United Service Magazine*, June, 1898.

colonies properly so called, the success of the British is thus explained by this authority :—

"The reason appears to lie in two traits of natural character. The English colonist naturally and readily settles down in his new country, identifies his interest with it, and, though keeping an affectionate remembrance for the home from which he came, has no restless eagerness to return. In the second place the Englishman at once and instinctively seeks to develop the resources of the country in the broadest sense. In the former particular he differs from the French, who were ever looking back longingly to the delights of their pleasant land, in the latter from the Spaniard, whose range of interest and ambition was too narrow for the full evolution of the possibilities of a new country."

The British colonies are protected in their infancy by the fleets of the mother country, and when they have attained their majority are well assured that, as long as sea-communications remain open, the mother country will purchase their wares, in state ceremonials do honour to their representatives, and in her centres of military activity gladly train their sons in the art of war.

The meaning of "command of the sea" in a strategic sense is summarised in the orders given to the Red and Blue Fleets during the naval manœuvres of 1893. "If the Blue side has either been defeated or has been compelled to retire to a distance to avoid an engagement, and the Blue torpedo vessels have been destroyed or reduced to inactivity, the Admiral of the Red side is to report by telegraph if he considers that his side has gained the command of the sea so that a large expedition can be sent across it."

The true method of protecting our islands from being either harassed by blockading-squadrons or invaded is not to keep a defensive force at home, but to hover around the enemy's ports and naval arsenals, as was clearly set forth by Drake, Raleigh, and Howard in the days of Elizabeth, and their theory was ably put in practice by Lords St Vincent and Nelson more than two centuries later. The effective strategy of viewing the enemy's fleet and attacking it was pressed upon Elizabeth by all her leading naval experts, but rejected in favour of awaiting the Armada in the Channel. The result might have been disastrous, but that the strategy of the Spanish was worse still, and that they were thoroughly defeated in tactics.

There are three kinds of war for our own country. There is first the war in which Great Britain begins with the command of the sea, and keeps it ; secondly, the war in which, not holding the command of the sea at first, she eventually acquires it ; and thirdly, the war in which the enemy ends by gaining the command of the sea. The third kind of war means the destruction of the British Empire, and, if the enemy wishes, the conquest of Great Britain. The second kind of war would put the British Empire into a condition of temporary dissolution for a longer or shorter time, the duration of which cannot be determined in advance, and would bring irreparable loss to British trade. The first kind, and the first alone, would secure the continuous maintenance of the Empire and of British trade, and the possibility of the further expansion of both[1].

The notion that our wealth afloat would be saved in

[1] *Command of the Sea*, by Mr Spencer Wilkinson.

a maritime war by the transference of our ships to a neutral flag and of our cargoes to neutral vessels has been based upon the supposed efficiency of the Declaration of Paris, 1856.

This notion is based on the idea that international arrangements have a sanction independent of force. A powerful enemy would soon put an end not only to such a transference, but also to the navy and shipping of the neutral State. Moreover, as we own more than 50 p.c. of the mercantile marine of the world, no neutrals have a sufficient amount of spare shipping to carry our cargoes as well as their own, whatever amount of freight our people might be willing to pay. To rely upon allies or neutrals is the desire of weakness. "*Nemo me impune lacessit*" is the motto of strength. At no critical period of natural development did any people find any effective assistance outside their own resources skilfully organised.

Bacon's doctrine is supported by the history of Japan, which has not been conquered for more than 1500 years, while she has been able to worry China for centuries, and, acting from a sea-base through the peninsula of Corea, to capture Port Arthur and Wei-hai-wei.

The Mexican Gulf and the Caribbean Sea are destined to be as important in regard to the strategic situation of the West Indies, North America, and South America as is the Mediterranean in regard to Europe, Asia, and Africa. Columbus, in search of Asia, came across the West India Islands, and going still westward, sought in vain all along the coast, from Cape Gracias à Dios to Cartagena, for some strait whereby he could attain India. For a while the Spanish had undisputed sway in these seas, but from the day when Cromwell's

Admiral, Penn, seized Jamaica to the epoch of Napoleon, the various European Powers struggled fiercely
with each other for the possession of the several
islands, which frequently changed hands. For a few
generations the Buccaneers made British and French
islands bases of operations against the Spanish traders
and the various towns on the coasts of these seas. They
recognised clearly the importance of the Isthmus, as
clearly as Capt. Mahan himself, and from 1625 to 1688,
having bases at Tortugas, Jamaica, and Old and New
Providence Islands, they harried and robbed again and
again Campeche, Vera Cruz, Maracaibo, Porto Bello, Cartagena, and all the other towns on the littoral and villages
in the interior. Pierre le Grand, Davis, Mansfield, L'Ollon-
nais, Grammont, de Graaf, Teach, Roberts, and Kidd were
no mere robbers ; they were maritime tacticians of great
ability, and they anticipated the operations of the regular commanders, whose battles in the wars from 1669 to
1802 made the West Indies so famous. The greatest
was the British pirate Morgan ; and his march across the
Isthmus with 1200 men in nine days from Chagres to
Panama, his sack of the town, and return to his base
were displays of rare military genius. The neglect of
the West Indies in recent years by our rulers is in-
explicable, inasmuch as, during our desperate strife with
Napoleon, they were the chief support of the commercial strength and credit that alone carried us to the
triumphant end. The Isthmus and the Caribbean Sea
were then vital elements in the constitution of our
Empire, and our trade with that part of the world was
many times more valuable than that with the Mediterranean. Moreover, the products of the Philippines were

conveyed to the Isthmus, and thence to Europe. It is most desirable that the attention both of statesmen and the commercial classes should now, as at the close of the 18th century, be directed to these seas, for, beyond a doubt, when once the canal across the Isthmus is made, whether its mouth be at Greytown or Colon, whether its course by Nicaragua or by Panama, the various islands will become trading and strategic positions of the very first order, and if we neglect them we shall lose no small portion of our naval and mercantile pre-eminence.

The two most important islands are Cuba and Jamaica. With regard to the former, its possession and a good fleet give command of the Mexican Gulf. Havana was once ours, and had not the peace-at-any-price party under Lord Bute been in power in 1763, and had we retained it, how different would have been the history of the island! The United States have it now, and Anglo-Saxons have superseded our ancient Spanish enemy. As to Jamaica, its capture in 1655 was a lucky accident, but Cromwell sought for good fortune by the right road of sea power. Hallam says :—" When Cromwell declared against Spain and attacked her West Indian possessions, there was little pretence of justice." Perhaps as much as in the recent American attack on Cuba, but in both cases the expeditions were most expedient and profitable. " So auspicious was his star, that the very failure of that expedition obtained a more advantageous position for England than all the triumphs of former kings." This was Jamaica. On the question of these islands and the Isthmus, one cannot do better than quote Captain Mahan, who says :—

"Wherever situated, whether at Panama or at Nicaragua, the fundamental meaning of the canal will be
that it advances by thousands of miles the frontiers of
European civilization in general, and of the United
States in particular; that it knits together the whole
system of American states enjoying that civilization as
in no other way they can be bound. In the Caribbean
Archipelago—the very domain of sea power, if ever
region could be called so—are the natural home and
centre of those influences by which such a maritime
highway as a canal must be controlled, even as the
control of the Suez Canal rests in the Mediterranean.
Hawaii, too, is an outpost of the canal, as surely as Aden
or Malta is of Suez; or as Malta was of India in the
days long before the canal, when Nelson proclaimed
that from that point of view chiefly was it important to
Great Britain. In the cluster of island-fortresses of the
Caribbean is one of the greatest of the nerve-centres of
the whole body of European civilization[1]."

The principal sea-routes in this part of the world are
as follows. From New Orleans by the Yucatan passage
to Colon. From New Orleans by Key West between
Florida and Havana, and between Florida and the
Bahamas to New York. From Colon by Kingston in
Jamaica by the Windward passage between Cuba and
Haiti to New York. From Cuba by the Anagada
passage past the island of St Thomas to Europe. The
Mona passage between San Domingo and Puerto Rico
is also of importance.

It is well to set forth clearly that no number of
forts, harbours, fortified arsenals, or coaling-stations give

[1] *Interest of America in Sea-Power*, p. 260.

command of the sea.   They are all valuable, as are also
torpedoes in harbours, and mines, whether as bases or as
means of protecting bases, or warding off bombardment,
or destroying blockading forces.   Too many naval
positions, like too many inland fortresses, are a waste of
resources.   With regard to coaling-stations, they are now
as essential to the life of navies as a supply of food is to
the life of man.   And the existence of coaling-stations
for an enemy's fleet within easy reach of a country is one
of the most serious threats to that country's security.
The Monroe doctrine forbade any European State to
make any further settlements in the American continent.
To this prohibition is now suggested a further restriction,
that no foreign State is henceforth to acquire a coaling-
station within 3000 miles of San Francisco, a distance
which includes the Hawaiian and Galapagos islands, and
the coast of Central America.   Hence the United States
has annexed Hawaii.   But no elaboration of defensive
positions, no coast defences, however strong, give command of the sea, though they are both valuable and
necessary.   Command of the sea depends on the sea-
going fleet, even as success in a land campaign depends
on the field army.   A strong fleet, capable of holding its
own against any hostile fleet, and of taking the offensive,
is the best line of defence for an island, and a not less
admirable defensive instrument for a continental Power
with a sea-frontier.   Under such conditions as the
modern competition and outward expansion of European
Powers and of the United States, a strong navy is essential for any community that would hold its own.   "The
command of both the Indies," is now, more even than in
Bacon's time, in the hands of the State that maintains a

strong naval force in a condition of constant readiness for action. Spain allowed her navy to be shut up in harbours and to be destroyed in 1898; the result has been the loss of the " Pearl of the Antilles " in the West Indies, and the Philippines in the East.

In connection with the command of the sea is the question of the command of navigable rivers. Once the sea power of an assailant becomes irresistible these are at his mercy. Thus the Federal gunboats used the Mississippi and the James Rivers as avenues into the heart of the Confederacy ; and thus British maritime supremacy is not limited by the coasts of the great oceans, it has ramifications into the centre of North America by the St Lawrence, it is felt at Mandalay and Benares, and by the banks of the Atbara and the Niger.

CHAPTER V.

THE STRATEGIC IMPORTANCE OF THE MEDITERRANEAN SEA.

NOT only has the Mediterranean been the birthplace of European commerce, refinement, and culture, but in every period of history the command of this sea has had a decisive influence on the strategy of the nations whose territories are on its shores, and also on the commercial and military policy of states abutting on the Atlantic, the German Ocean, and the Black Sea. The strategic geography of the islands and positions on its shores would require a bulky volume, but some of the most important of its features can here be indicated.

Large as the sea is—2300 miles in length, and in places nearly 500 miles wide—its avenues of entrance are narrow. The Straits of Gibraltar are not quite nine miles wide, the Dardanelles and the Bosporus each about a mile. Forts mounted with the best modern guns would make the entrance of a hostile fleet either into the Mediterranean or Black Sea a dangerous experiment. By its shores or on its waters the fate of Europe and Asia has often been decided.

The battle of Salamis saved Europe from an Asiatic domination. The later records of Greece and Syracuse, and the realms governed by the successors of Alexander, abound in striking examples of the importance of naval actions and naval bases in this sea. The wars between the Romans and the Carthaginians were at first chiefly maritime, and though Hannibal, by the possession of Spain, was able to make a march overland to the very south of the Italian peninsula, he was obliged to return for the defence of Carthage when Scipio, having the command of the sea, transferred the war to Africa. Even in the civil wars of Rome, as Pompey said, following the counsel of Themistocles,—"He holds that whoever becomes master of the sea, becomes master of all things." Pompey would have worn out Cæsar if he had kept to a maritime policy in the Grecian seas, instead of risking all on the battle-field of Pharsalia, 48 B.C. The loss of the battle of Actium ruined Mark Antony, 31 B.C.

The importance of the Aegean Isles at large, and of such greater islands as Cyprus, Rhodes, and Crete, was of the first order in ancient times, and till the realms of Antiochus and Pyrrhus fell under the sway of Rome. For a long period the people of Italy relied on Egypt as we rely on America for corn, but when the Vandals won command of the sea under Genseric this source of supply was cut off. The merciless and astute Goth had worse things still in store for the capital of the world. He had passed from Spain into Africa and annexed Numidia, Mauritania, Carthage, Egypt, Asia Minor, and Thrace. He then, in 455 A.D., embarked a mighty force at Carthage for Rome itself. He sacked the city for fourteen days, and thus, after more than six centuries, Hannibal was avenged.

After the close of the dark ages Genoa, Pisa, and
Venice became the centres from which the luxuries of
the East were transmitted to the cities of the Hanseatic
League, and thence to western and northern Europe.
The territories subject to Venice when it had " married
the Adriatic" and surpassed all Italian rivals, included
stations all along the Adriatic and positions in the
principal islands.   Powerfully supported by the Knights
of St John, the Venetians were the bulwarks of Europe
against the Turk.   No battle in any age was more
fateful than Lepanto.   The Turks had entire control of
the Levant, they had a splendid fleet, skilfully manned.
But Don John of Austria, commanding the Venetians,
defeated them there in 1571, and they were confined to
the western Asiatic and Grecian shores till at Navarino
in 1827 Christendom became in turn the aggressor.

When America and the route to India by the Cape
of Good Hope were discovered at the end of the 15th
century, the centre of commercial, and consequently of
strategic gravity, was transferred from Italy to the Iberian
Peninsula.   After the revolt of the Dutch subjects of
Philip II. of Spain and the naval success of Elizabeth
of England, "westward the course of Empire took its
way."   Holland and England played to Europe in the
17th century the parts enacted by Genoa and Venice
in the 14th and 15th.

Rhodes was the last of the bulwarks of Christendom
in the East to surrender to the masterful Turks ; it held
out for seventy years after Constantinople had fallen.
It was once, as its name implies, "The Rose of the
Levant."   During the Peloponnesian War, and the other
struggles that marked the decline of the states of old

Greece, it evenly balanced itself between the contending parties, always siding with the stronger. The heroic resistance of its capital to Demetrius Poliorcetes, who applied to its attack all the engineering resources of his time (B.C. 304), gave to the island the glory of war in addition to its previous fame for beauty and archi- tectural splendour. It was plundered by Cassius, and thenceforward became alternately free and vassal under successive emperors. The Saracens are said to have sold its fallen Colossus for old metal. The Knights of St John received charge of the isle from the Byzantine emperor in 1308. They were past masters in the art of fortification, beating off Mahommed II. in 1480, and resisting enormous numbers of the Turks for four months in 1522, at a period when the Moslem chiefs were all Ghazis. The Turks allowed De Lisle Adam, the' Grand Master, to transfer his heroes and himself to the island of Malta, where they forthwith began to erect a series of beautiful and imposing works named after the leading nations of Europe, from which the brilliant and enter- prising Knights were recruited.

Malta has since had a most interesting history. Our people have in this age no idea of the horror with which the Moorish and Turkish corsairs of the Mediterranean were regarded in the days of Elizabeth and Charles I. Cromwell sent an expedition against the pirates of Algiers, but they played havoc with commerce, and even exacted tribute from our own Government, till their place of arms at Algiers was taken by Pellew in 1816. The Knights of St John were safe behind their impregnable fortifications ; the whole isle was a fortress, with stations all along the coast and military roads in every direction. From the point of view of defence it was, indeed, more than equal to the requirements of the time, and considering the interests at stake far superior to its present condition.

The Moslem corsairs, such as Byron celebrates in 'Don Juan,' were always at issue with the Knights. Their redoubtable chief Dragut, the pupil of Barbarossa and the rival in seamanship of Andrea Doria, ravaged Gozo but was killed at Malta in 1565. The Turkish admiral was reinforced by all his co-religionists in the Mediterranean, and maintained a series of able and desperate assaults from May 18th till September 7th. The skill both naval and military displayed on both sides was seldom equalled, but 9000 of a garrison repulsed 40,000 assailants. Within their works, which they kept on strengthening till Fort Tigné was finished in 1793, the Knights erected that lovely shrine to their patron saint which Sir Walter Scott considered to be one of the finest in Europe, but they also took care to construct in honour of the Grand Master Valette, who repulsed the Turks, the beautiful town and safe harbour of Valetta—one of the few towns " built by gentlemen for gentlemen."

The Moslem energy once exhausted, however, the Knights seem to have sunk into a state of apathy. Having been the safeguard of Christendom in their mid-ocean fortress, and having won again and again the thanks of all Christian nations, and the hearty admiration of their high-spirited adversaries, ease and luxury began to take the place of activity and skill. Their immediate surrender to the summons of Bonaparte, who could not possibly have spared time either to assault or invest the smallest of their positions, is another

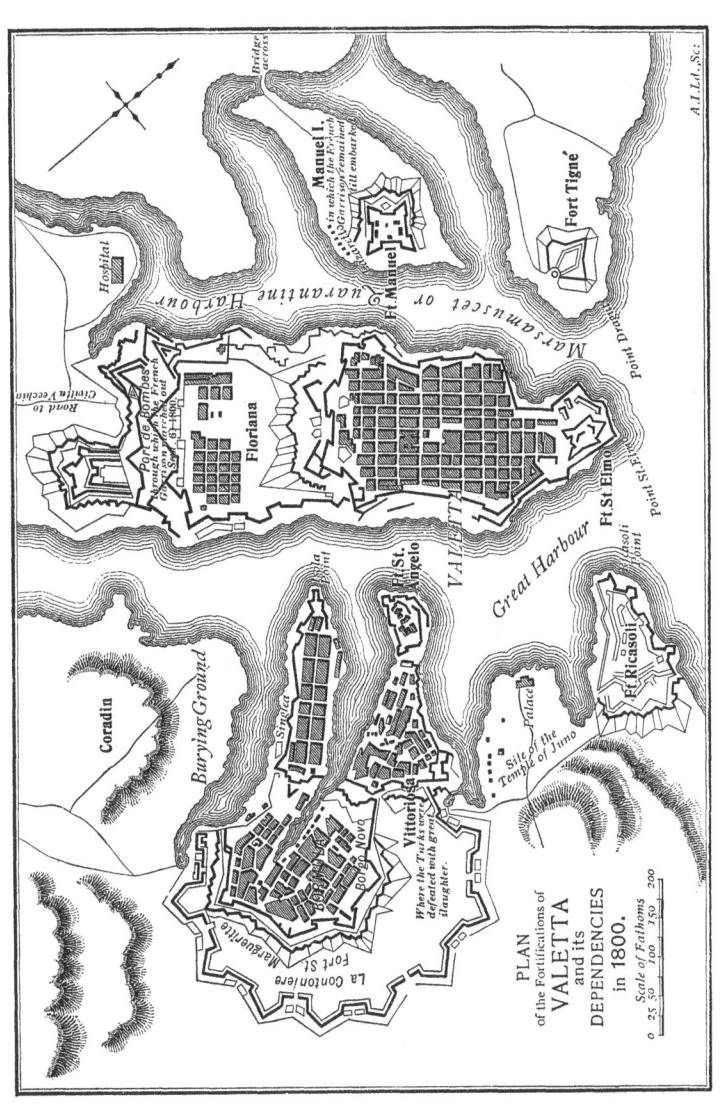

PLAN
of the Fortifications of
VALETTA
and its
DEPENDENCIES
in 1800.

Scale of Fathoms
0 25 50 100 150 200

A.J.Ld., Sc.

Fort Tigné

Manuel I.
in which the French
Garrison remained
till embarked

Ft. Manuel

Marsamuscet or Quarantine Harbour

Point Dragut

Hospital

Road to Citatta Vecchio

Floriana

Part of the
French Ground
Garrison marched and
Surrendered
1800

VALETTA

Point St. E.

Ft. St. Elmo

Great Harbour

Bridge across

Point Dragut

Coradin

Burying Ground

Senta

Ft. St. Angelo

Site of the
Temple of Juno

Palace

Ricasoli
Point

Ft. Ricasoli

Marguerite

Fort St.

La Contoniere

BORGO NOVO

Vittoriosa

Where the Turks were
defeated with great
slaughter.

illustration of the axiom that the spirits and muscles of men, and not guns or bastions, are the defence of nations. There was not even a bombardment. Alison and the French general Caffarelli suggest corruption on the part of the French Knights; if so, they ruined their famous Order in June, 1798. The good fortune of Bonaparte was the more remarkable, inasmuch as the British, with absolute command of the sea, found some difficulty in taking the fortress, although it could not possibly be relieved from outside, whereas if the Knights had resisted *à outrance* in 1798 Nelson would have come to their assistance. The French Commandant Vaubois, with 6,000 men, withstood a siege of two years by Maltese, Neapolitans, Portuguese, and English, but he surrendered in 1800, and the island became a British possession. This was one of the causes of the renewal of the war after the Peace of Amiens. Napoleon said, " I would rather have the English on the heights of Montmartre than in Malta." In future complications in which the Orient is either combined with, or adverse to, the West, Malta will play a leading part. Moslem and Hindu soldiers from the banks of the Ganges and the Indus were quartered in the old palaces and casemates of the Hospitallers in 1878. Nor could the ablest of the Grand Masters have conceived such a display of fighting power and commercial prosperity as is in this generation constantly entering and leaving the harbour of Valetta under the standard of St George.

In 1669 the Venetians lost the island of Crete, which was taken by the Turks after a siege continued at intervals for twenty-five years, during which the besiegers lost 120,000, and the Republic 30,000 men.

Syracuse, though now of little strategic importance, and of no influence as a fortress since its destruction by the Saracens in the 8th century, once played a leading part in the military history of the Mediterranean. Its siege by Nicias was one of the most striking and celebrated incidents of the Peloponnesian War, and the study of its details is an epitome of ancient military and naval tactics. When the remnant of the Athenian army was shut up to perish in its quarries, their country had begun its career of inglorious decay. The city which Pericles had adorned sank within a few generations into a secondary position simply because of democratic turbulence and an inefficient general. Nicias, we are told, was the most respectable man of his age, but in war a vigorous character is worth more than all the private virtues. Nicias brought about the ruin of one of the most powerful expeditions of Grecian history 413 B.C., while the unscrupulous and cruel Dionysius made Syracuse, after it had attained military renown, one of the most artistic, wealthy, and powerful cities of the Grecian world. In regard to the Athenian siege of Syracuse the adage "what is wanted is not men but a man" applies with force. In 1814 Napoleon said,— "I have 50,000 men and myself, and that makes 150,000 men." And he was right: his adversary Wellington said the mere presence of Napoleon with an army was worth 40,000 men.

Sicily naturally occupied a prominent strategic position in the long struggle between Carthage and Rome for the command of the sea. When in the Second Punic War the Syracusans were so ill-advised as to take the side of Carthage they sealed the doom of their city as an independent or free state. The siege by the Romans is a familiar story in all our schools, far more carefully studied than the siege of Badajoz. Not even the formidable engines of Archimedes could long avert the sack of Syracuse.

That singularly active Norman race which conquered England occupied Sicily in 1266; the phrase "Sicilian Vespers" still preserves the memory of the massacre by which the inhabitants got rid of them in 1282.

The British during the Napoleonic wars not only occupied Sicily and preserved it for the Bourbons when Joseph and Murat ruled in Naples, but also used it as a base for combined military and naval operations in Calabria and Catalonia.

In the centre of the Sicilian Strait is the island of Pantelleria, belonging to Italy, but it is not fortified.

In the Roman wars the services of the skilful slingers from the Balearic Islands were in constant request. Port Mahon in Minorca was the best harbour in the Mediterranean in the opinion of Andrea Doria. Stanhope seized it for the British in 1708. It fell to the French 1756, and was restored, but it was finally taken from us in 1782.

The settlements along the coast of the Levant from old Troy to Acre have stood the brunt of many a siege or naval attack; the most interesting to us, perhaps, are those of the latter city. Three times at critical historic epochs have British forces appeared before it. Richard Cœur-de-Lion took it in spite of Saladin; Sir Sidney Smith assisted in its successful defence against all the efforts of Bonaparte in 1799; and our fleet by the bombardment of 1840 stopped the march of Ibrahim Pasha against the Anatolian provinces of Turkey, and perhaps against Constantinople itself.

Corfu was a fine British fortress up to 1864, when it was given up to Greece for sentimental reasons. Yet its surrender would not have seemed a wise policy to Napoleon, who said, "San Pietro [S.W. of Sardinia], Corfu, and Malta will make us masters of the whole Mediterranean."

The British siege of Toulon in 1794, when Bonaparte made his first successful strategic suggestion ; the operation about Barcelona in 1704–5, where Peterborough displayed romantic genius ; the capture of Gibraltar, 1704, and its subsequent sieges, especially its brilliant defence by Elliot against all the resources of France and Spain under de Crillon, are part of the schoolboy history of our Empire. But the northern coast of Africa has also felt the force of British naval and military skill. Thus, Pellew destroyed the power of the corsairs of Algiers in 1816. The battle of the Nile, 1798, was not so much a fight as a conquest. By Abercromby's victory at Alexandria, 1801, and the surrender of Menou, Bonaparte's Egyptian expedition was brought to an inglorious close. The bombardment of Alexandria in 1882 was the first test of the power of modern ships of war and their guns against forts ; and the subsequent burning of the city and other proceedings of the Arabs were proofs of the futility of a mere naval attack unaccompanied by a military force able to land at once and to act with efficiency.

The position of Byzantium justified its choice by Constantine as the site of the capital of the Eastern portion of the Roman Empire 330 A.D. History has during 1550 years borne this out. Throughout all this period the city has been a leading object in the ambitious dreams of every conquering race, and a strategic centre of the first importance. Its future fate is even now a matter for most anxious thought in every European Court. It has every advantage from the commercial and military point of view. From her seven hills the imperial city commands the shores of Europe and of Asia. Gibbon's glowing description is not exaggerated, as every traveller who has passed from the Golden Horn to the Black Sea can bear witness. Since the days of this historian a feature of melancholy interest has been added for the British traveller; for are not the graves of many a British soldier at Scutari on the right? The climate is healthy, the soil fertile, the harbour excellent, the approach by land narrow and easily defended. When the gates of the Hellespont and Bosporus were shut, ample supplies of bread, wine, and meat could be procured from the coasts of the Sea of Marmora, and fish abounded in the surrounding waters. When the straits were open, the riches of all the known world were poured into the storehouses and bazaars of the celebrated capital, which age after age was the bulwark of Christendom against barbarism. The ruler of Constantinople, as long as he held the straits, had as much sea power as ancient times could provide.

From the very beginning the city was of preeminent utility. The barbarians of the Black Sea in the generation preceding Constantine had ravaged the Mediterranean coasts after the fashion of the Vikings in the British seas at a later date, but thenceforward their incursions ceased. The attacks on Constantinople began

in the middle of the fifth century and continued till 1453.
It was only captured three times. In 450 the Huns
assailed it; in 553 the Huns and Slavs; in 626 the
Persians. It was besieged by the Arabs seven times
between 668 and 782, by the Russians four times,
but unsuccessfully, from 865 to 1043, and by the Hungarians in 924. Its first great catastrophe was its
siege and capture by the Crusaders and Venetians, who
sacked it in 1204, burned portions of the town, destroyed
many famous works of art, and the altar of St Sophia,
and possessed themselves of a spoil equal in value to
seven times the whole revenue of England at that period.
Michael Palæologus drove out the Latins with the aid
of Varangians in 1260, and the Sultan Amurath vainly
attempted its capture in 1422. It was soon to fall into
Turkish hands, however, for Mahommed II. carried it in
1453 after a gallant defence of fifty-three days by
Constantine Palæologus and the Genoese Giustiniani.
The surprise of the unguarded gate of Kerkaporta ought
to be a warning to the defenders of fortresses to leave
nothing to chance.

A very regular system of *passagia* from the towns of
the Mediterranean coast to the Holy Land was arranged
throughout the Crusades. The exact dates cannot be
ascertained, but there was a spring passage (*passagium
Martii*) and an autumn passage (*passagium Augusti*).
Richard Cœur de Lion's main fleet started from the
different harbours of England, Normandy, Brittany, and
Poitou after Easter, March 25th, 1190. They coasted
Brittany, Poitou, and Gascony, but when rounding Spain
were severely handled by the weather, though they
reached Lisbon in safety. They left it July 24th, and

8—2

being reinforced, passed the Straits August 1st. Coasting Spain by Tarragona and Barcelona, they reached Marseilles. Leaving this port August 7th, they coasted by Genoa, Pisa, and Naples. They crossed the Straits of Messina with 100 ships and 14 strong two- or three-masted "busses." Each ship—in addition to 40 soldiers, 15 sailors, and 40 horses—carried a year's provisions for man and horse.

The busses took double cargo and gear. The treasure was divided among the ships. Messina was safely reached and taken, and the Greek or "Griffin" robbers and extortioners mercilessly punished by the "English Lion." It was here that the Crusading chiefs issued the elaborate regulations about commissariat and markets which are preserved in Howden's Chronicle. From Messina Richard went to Catania to interview Tancred, King of Sicily, having first met Philip of France, who settled all his difficulties with his fellow Crusader and gave him 10,000 silver marks. Philip left Messina in March, 1191, and reached Acre on the 21st April. Richard followed him April 10th, and after a very stormy passage reached Cyprus, and having reduced Isaac its emperor to submission, he married Berengaria at Limasol. When he had arranged his baggage, he set sail for Acre, but before leaving, appointed energetic men as his captains and wardens in Cyprus, leaving them instructions to send after him what victuals were necessary, namely, wheat, barley, and the flesh of all the animals in which Cyprus abounded. On his way to Acre, he took a great Saracen ship, the Dromond, which had many supplies for the Acre garrison; and it is probable that, had the ship got safe to its port, the Christians would never have taken

the city.   The ship contained seven Emirs, and 800
chosen warriors, 100 camel-loads of arms, heaps of bows,
spears, and arrows, a great stock of food, and much
Greek fire.   Saladin's confusion equalled that of Bonaparte's six centuries later, when Sir Sidney Smith performed a similar feat.   " Now have I lost Acre," he
exclaimed, " and besides those chosen men, in whom I
placed my trust ; I am overcome and oppressed by the
harshness of my fate."   How similar this action of
Richard's fleet is to that of Sir Sidney Smith's in 1799 !

Richard reached Acre June 8th, 1191.   His fleet
completed the investment, and the conquest of Cyprus
gave the assailants an excellent depôt of supply.   After
the capture of Acre he went along the coast to Cæsarea,
and thence marched to Arsuf, utterly routing the Saracen
host.   He reached Beit Nûba 13 miles N.W. of Jerusalem
in December 1191, and began to retire January 1192.

The first volume of Sir W. Hunter's admirable
History of British India recently published makes it
so very clear that the ancient lines of communication
corresponded on the whole with the modern, that some
excerpta from it are desirable, especially as the position
of Egypt now occupies such a prominent place in the
policy of our people.   A side issue of some interest to
economists is that the Romans seem to have been as
much concerned about the excess in money value of their
imports and their exports as some of our own politicians;
in their time the balance of exchange was distinctly in
favour of India.   From the founding of Alexandria
(332 B.C.) its Asiatic trade grew with the improvements
in the sea-passage.   At a very early period the Arab
navigators tried to avoid the northerly winds which

sweep down the Egyptian coast, by unlading their cargoes near the modern Kosseir, and transporting them overland to Thebes, the capital of the Nile valley. Ptolemy Philadelphus did much during his long reign (285—247 B.C.) to concentrate the Eastern trade at Alexandria, the new capital of Lower Egypt. He re-opened the ancient cutting from Bubastis to the Bitter Lakes, and was only stayed from completing his canal to the Gulf of Suez by fears lest the Red Sea would flow in and submerge the delta. To escape the difficult naviga-tion of the Suez Gulf, he founded, on the headland near its mouth, Myos Hormos (274 B.C.), whence the Indian wares were carried across the desert to the Nile valley. Still further to avoid the northerly head-winds on the passage up the African coast, Ptolemy created the emporium of Berenice at the southern extremity of Egypt on the Red Sea, and honoured it by his mother's name. A caravan journey of 285 Roman miles con-veyed the eastern freights across wastes and mountains to Coptos on the Nile, with regular halting stations along the tract. Some of these still dot the desert, and the proposed Assuan-Berenice railway—for which a survey is at the moment of writing these lines being made— would revive the old trade-route from Ptolemy's harbour to the Nile valley by a shorter cut. Railway communi-cation seems destined, indeed, to reopen the paths of Indo-European commerce. The Russian line to Bokhara represents, not too exactly, an old route from China by way of the Oxus, and the long-projected Euphrates Valley Railway would be the modern counterpart of the Syrian caravan track. The final development of the Indo-Egyptian route did not take place until three

centuries after Ptolemy Philadelphus, when the pilot
Hippalus discovered the monsoons, or more strictly
speaking, worked out the regular passage by means of
them (circ. 47 A.D.). While, moreover, the Egyptian
coast-passage is impeded by northerly winds during the
greater part of the year in the upper part of the Red Sea,
the navigation at its southern end is aided by regular
alternations in the air currents, southerly winds predominating from October to June, and northerly winds from
June to October. The establishment of the emporium
at Berenice in the third century B.C. thus paved the way
for a vast expansion of the Eastern trade as soon as the
monsoons were put to their mercantile uses in the first
century A.D.

Egyptian merchant fleets sailed from Berenice or
Myos Hormos in July, rounded the modern Aden with a
halt at Kanê in August, and were blown rudely across
the Arabian Sea to Malabar by the middle of September
—a voyage of sixty or seventy days from the Egyptian
to the Indian coast. Having sold their western freights
and bartered their bullion for eastern cargoes, they
started from India at the end of December, and were
wafted more gently back by the monsoons to their Red
Sea harbours about the beginning of March. This
monsoon route became the chief channel for the bulkier
produce, as well as for the precious gems and wares of
India ; it enriched the ports along its line, and made
Alexandria the commercial metropolis of the Roman
Empire. Pliny lamented the vast shipments of gold and
silver sent from Europe to pay for the products of Asia.
" In no year," says he, " does India drain our Empire of
less than fifty-five million of sesterces (£458,000), giving

back her own wares in exchange, which are sold at one hundred times their prime cost."

Of this great commerce, while Egypt still remained a Roman prefecture, two accounts by actual traders exist. The "Periplus of the Erythræan Sea," or the circum-navigation of the Indian Ocean, as it may be rendered, describes it within a hundred years after the discovery of the monsoon winds by the pilot Hippalus. Written probably by a Greek merchant who had settled at the southern Red Sea emporium of Berenice and voyaged to the East, its composition is assigned to some time between 80 and 161 A.D. It gives the seaports on the route, specifies ninety-five of the chief articles of traffic, and forms a wonderfully complete presentment of the Indo-Egyptian trade in the first century of our era.

The permanent importance of strategic points, whether sea power or land power be concerned, is manifest from a study of Mediterranean depôts, arsenals, and routes, both in the sea itself and radiating from its ports inland.

Going round its shores it will be found that every locality which the ancients felt to be of first importance has practically the same strategic influence now. The "Pillars of Hercules" in Europe and Africa will ever be jealously watched by rival powers, Carthagena is of interest to Frenchmen as being the nearest *entrepôt* to Africa, Biserta takes the place of the port and fortress of Carthage, for Tunis is kept by the French in spite of treaty phrases. The towns on the Syrian coast have had the same strategic and commercial importance ever since the time of the Phœnicians; the position of Delos still makes it the centre of the Grecian world; the islands and harbours near the Troad have been in this century the rendesvous of mighty naval squadrons, and would assuredly be an object for any western naval Power desiring to conquer what is left of Turkey in Europe or to command the Black Sea. The trade of Salonika is rapidly reviving; had the Greek navy been able to seize it or Dedeagatch near Enos Bay, the Turkish operations of 1897 would have been crippled at the start.

In olden times, when roads from ports into the interior of countries were bad, it was the object of merchants to get up to the very top of gulfs and bays, if feasible, in order to shorten the land journey as much as possible. Now that railway communication is preferred to the delay of a long sea voyage, the nearest port at the apex of a peninsula, as Brindisi, supplants ports like Venice or Genoa, and canals to reduce long sea distances are in vogue. The Suez and Corinth canals and the proposed ship-canal by the Bay of Biscay to the Gulf of Lyons are examples, as is the canal from the Baltic to the Black Sea which the Russians have in view.

The reports and rumours during the last year concerning the construction of this latter great waterway, though very conflicting, still lead one to suppose that it is feasible, and has been seriously contemplated, even if the work has not proceeded very far. The route proposed—as already mentioned—is from the Gulf of Riga, by the rivers Duna, Beresina, and Dnieper, to Kherson on the Black Sea, and fifteen ports or harbours are to be constructed at various places situated along its whole course of 994 miles. The channels of the rivers are to be deepened and new cuttings made where necessary, so as to give a minimum depth of 28 feet of water.   It is estimated to cost £20,000,000 sterling, or about £5,000,000 less than the amount said to be required for cutting the Nicaragua Canal, and it will take five years to complete. The primary object of this great undertaking is to connect the naval dockyards at Libau with those at Nicolaieff, and permit of the passage of Russian men-of-war to and from the Black Sea and the Baltic, thus neutralising to some extent the closing of the Bosporus and Dardanelles in time of war.   The transit from sea to sea will take six days.   Moreover there is little doubt that such a ship-canal passing through Russian territory from end to end, developing a very rich tract of country, and bringing sea-borne traffic to the very gates of what have hitherto been inland towns, will prove a very great advantage to the trade of the country, and is bound to be a commercial success, while the physical features of the land and especially the existence of a clay soil throughout its whole length are very favourable to its construction.   According to a usually reliable authority, it is estimated that about one-eighth of the canal only will have to be wholly artificial, and that only two locks will be needed.   The worst difficulties will arise about the upper portion of the Dnieper, where it flows through marshy forests, and 200 miles from the mouth of this river there are a series of nine rapids falling 107 feet in 40 miles.   The town of Ekaterinoslav on the Dnieper is 161 feet above the sea-level, while Alexandrovsk about 50 miles to the south, on the same stream, has an elevation of only 49 feet.

Arsenals and naval depôts as contrasted with commercial centres are placed as high and as near the

continental portions of peninsulas as possible; for example, Pola, Ancona, Spezia.

The towns in the south of France, Marseilles and
Toulon, as means of attack are very useful. Thence
have issued many excursions against other European
States, but though they more than retain their ancient
value, they are not likely to be attacked. The failures
of the British in 1706 and 1794 demonstrate this, and
though Nelson watched Toulon he did not attack it. It
was almost neutralised by our possession of Port Mahon,
and, as to eastern expeditions, it is seriously affected by
Malta and Cyprus.

The proposed abandonment of the Mediterranean in
1796 was one of the most egregious blunders in our
history, but the policy of withdrawal and limitations of
our liability has always advocates in our midst. Some
writers would have us abandon our strategic position in
this sea, and allow it to become a French lake. The
commercial and military objections to this proposal are
set forth by Sir George Clarke. It would be an extraordinary and utterly profitless act of self-abnegation on
our part, and would at once reduce our lines of communications between East and West from interior to
exterior lines, a policy as fatal to strategy on sea as on
land. The future is of course uncertain, but the value
of sea power in the Mediterranean is writ too large in
history for any wise politician to venture to risk the
consequence of its loss. National honour, splendid
traditions, and the eternal principles of naval strategy
alike forbid us to desert our commerce, and that of our
colonies, on 3,000 miles of the element which we have
been taught by successive generations of sea-officers to

call our own.   If we abandon the Mediterranean and
hand over to our rival the spoils of a great naval victory
without firing a shot, we give to the world the sure sign
of that madness which the affairs of men and of nations
on the downward road preface with[1].   In short we are in
the Mediterranean, (1) because history shows that we
must be there, (2) because our commerce there afloat
is enormously greater than that of any other Power,
(3) because no other waters would serve equally well
for the training of our fleet.   In the Mediterranean we
must remain if our Empire is to last.

But it must once more be reïterated that no fortress,
no harbour, no multitude of harbours can give sea power,
which depends alone upon an efficient fleet.   To this
fortresses and harbours are undoubtedly useful hand-
maids[2], but they cannot supersede it, and without it they
are merely inert, though when well placed and supplied
with modern works and guns they could defy a fleet or
seriously hamper its offensive power.   Expeditions must
have good harbours for their transports, and if the
Spanish works in Cuba and Manila had been furnished
with trained men and with proper materials the American
Admirals would have had a difficult task in 1898.   Our
fleet also did not seriously attack Rochefort or Brest in
1804–5, but they bottled up the squadrons of the enemy.
In the same way many competent authorities fear that
if the Toulon fleet became superior to our Mediterranean
fleet, and moved athwart the Straits of Gibraltar, our
vessels might be similarly pent up, cut off from succour,

[1] *The Navy and the Nation*, page 242.
[2] See Debates in Parliament April 14th, 1899, upon the question of
fortifying Wei-hai-wei.

and, unless ample stores were provided in Malta, reduced to submission in very short time. It is a question
if Malta, as a depôt, is up to modern requirements,
whether reserves of munitions of war or of provisions
be considered.

CHAPTER VI.

THE UNITED KINGDOM WITH REGARD TO SEA POWER.

THE military enterprises of the Middle Ages brought Britons into touch not only with Continental Europe but with Asia. The exploits of Richard I. and his knights in the Holy Land were prominent among the numerous adventures that illustrate the prowess of Christians and Moslems along the coasts of Asia Minor, Syria, and Africa. Other Englishmen dared to join the Teutonic knights and repel the advances of the heathen of the north. The varied fields on which English valour had shone are pointed out by Chaucer in his *Canterbury Tales*, who says of his knight :—

> " At Alisaundre he was when it was wonne
>   Ful ofte tyme he hadde the bord bygonne,
> Aboven alle naciouns in Pruce.
>   In Lettowe hadde he reysed and in Ruce,
>   No cristen man so ofte of his degre.
>   In Gernade atte siege hadde he be
> Of Algesir, and riden in Belmarie.
>   At Lieys was he, and at Satalie,
> Whan they were wonne; and in the Greete see
>   At many a noble arive hadde he be.
> At mortal batailles hadde he ben fiftene,
>   And foughten for our faith at Tramassene."

The soldiers of the Black Prince and of Henry V. made many a famous expedition by the banks of the Somme, the Seine, the Loire, and the Garonne, and the free companies of Hawkwood passed into Italy, and were prominent champions of the various rich northern republics during their ruinous conflicts for commercial and political predominance. The Black Prince passed the Pyrenees and fought a great battle near Najara, while some of his knights had a desperate affray in 1367 on the very hill which Picton's third division carried with a rush during the battle of Vitoria in 1813.

While our soldiers and sailors were struggling against Spanish maritime power from Cadiz to the West Indies, and against the matchless infantry of Alva and of Parma in the Low Country, the brilliant writers of Elizabeth's time clearly appreciated the consequences to mankind of the discoveries of America and of the sea-route to India. They wrote that if Britons were wise the wealth of the world must be the dowry of their country. Bacon waxed enthusiastic about the future of British power by land and sea, and with philosophic acumen summarized the best policy to pursue with regard to the planting of colonies. Patriotism is the keynote of the historic plays of Shakespeare—he would, if need be, defy the shock of the four quarters of the world in arms. Sir Walter Raleigh not only trimmed and singed the King of Spain's beard with sword and fire, but in pregnant pages laid down the principles underlying the permanence of our Empire with a clearness which no modern writer has surpassed. In his *Faery Queen* the sage and serious poet Spenser allowed his imagination to foreshadow many a form of power and enterprise

then unknown, to which our later history has given a
local habitation and a name.

> " But let vain man with better sence advise
>   That of the world least part to us is red
> And daily how through hardy enterprise
>   Many great regions are discovered
> Which to late age were never mentioned.
>   Whoever heard of the Indian Peru?
> Or who in venturous vessel measured
>   The Amazon, huge river, now found trew?
> Or fruitfullest Virginia who did ever view?"

These lines were published two years after the fate
of the Spanish Armada gave our hardy sailors a fair
field on the waters of both the Eastern and the
Western Seas.

Let it not be supposed that the wind and the waves,
or luck, or any species of miracle came to the aid of the
British in 1588 any more than in 1805. Providence
never favours the incompetent or inert. True, when
Lord Howard and his coadjutors could no longer pursue
it, the winds battered what was left of the Invincible Armada against the dreary rocks of the north
of Scotland and the west of Ireland, even as another
gale wrecked the ships captured by Nelson. The
same qualities which made Napoleon and his corps the
masters of southern and northern Germany in 1805
and 1806 gave England the victory over the Spanish.
Our navy was superior in everything except bulk and
numbers. The English had been taught by Drake a
rational system of sailing-tactics and a sound marine
strategy. They won by science, by a more intelligent
comprehension of the art of naval construction and of
gun construction, they were better led, they shot faster,
and they carried more powerful batteries. Any defects

of their fleet were due to the politicians, and not to the naval instructors of the period.

But for some time the British had a vigorous competitor for supremacy, not only over distant Spanish settlements, but in their own seas. The Dutch, having won freedom from their Spanish masters, won wealth by their maritime energy. They dared to fling the gauntlet down to Cromwell when he was already master of England, Scotland, and Ireland, and when some of the ablest Royalist officers had joined his service. But, though gallantly and ably commanded by chiefs like Van Tromp and De Ruyter, the fleets of Holland were frequently defeated by Blake, Monk, Sandwich, and the Duke of York, and ultimately retired from the contest for maritime supremacy. Once the Dutch Stadtholder, William, became King of England, Holland and England were involved in a nine years' war against France, and Ireland was entered by French troops, leaving England in danger of invasion. But the battle of the Boyne, the indecisive battle off Beachy Head, and the decisive victory of La Hogue delivered our islands from all fear of invasion for more than fifty years.

In the reign of Queen Anne England became not only a sea power, but *the* sea power, and the navies of our new competitors in the paths of colonial expansion were regularly beaten in all parts of the world, from the battle of Malaga, 1704, to that of Trafalgar in 1805.

By the close of the War of the Spanish Succession we had already established trading-stations under the suzerainty of the Mogul emperors in Hindustan, while adventurers and pilgrims had planted our flag on the eastern coast of America from Massachusetts to Carolina.

Jamaica, St Helena, and the Guinea Coast had been secured. The Peace of Utrecht in 1713 brought into international prominence our colonial and mercantile policy.

In Europe throughout the 18th century we cham-
pioned the balance of power, whether threatened by
France, Austria, or Prussia, while our statesmen and
people strove might and main to nurture our commerce
by every fair means, and occasionally, it must be confessed, in the Spanish main, by methods which were
not always justifiable. Our fleets were kept up to a
wonderful standard of excellence, considering the wealth
and population of the period. Even so early as the
middle of the War of the Austrian Succession, 1740—
1748, the spirit which produced Thomson's "Rule Bri-
tannia" pervaded all classes of our people.

The celebrated cruise of Anson during this war was
a terrible blow to the prestige of Spain, and the victories
of France on land were neutralised by our naval power,
which saved our position, albeit not used to the best
advantage.

During the interval between this and the Seven
Years' War the schemes of Labourdonnais and Dupleix
in the East Indies failed because of our naval pre-
eminence, though Clive and his agents were 10,000 miles
away from their base of operations. Clive wrote to
Pitt:—" The superiority of our squadron, and the plenty
of money and supplies of all kinds with which our friends
in the Carnatic will be supplied from Bengal, while the
enemy are in such total want of everything, without
any redress, cannot fail of wholly effecting their ruin."

It would be a good exercise for a student to take
a map of the world and a dictionary of statistics, and

fairly estimate the difficulties in the way of such a complete mastery of strategy, and control of the avenues of communication with the most distant regions, as the British maintained from 1757 till 1762. There can be no question that supreme energy and genius were at work. Suffice it to say that they sent large forces to the country between the Weser and the Elbe, and that the British infantry won the admiration of their German leader Ferdinand, and their French enemy Contades, by the prowess they exhibited at Minden. By the banks of the Hooghly another British regiment, the 39th, *primus in Indis*, repulsed Surajah Dowlah. At the mouth of the Vilaine in France Admiral Hawke immortalised himself in 1759, and enabled the British fleet to do as it pleased between Europe and America. At Lagos Bay Boscawen ruined another French fleet in the same year. By the banks of the Ohio and on the coast of Cape Breton were laid at this period the foundations of Anglo-Saxon communities. The Americans are justly proud of their colossal manufactories at Pittsburg, but these works would have been French but for our desperate efforts at Fort Du Quesne. They have reaped the harvest sown with British toil on fields fertilised by British blood. Wolfe carried the heights of Abraham and handed down the Gibraltar of America to the British dominion of Canada; Amherst took Montreal, then a mere strategic point, but now one of the most important centres of the commerce of mankind. On the West African coast our cruisers protected our traders. In the Mediterranean our flag, which the unfortunate Byng had furled, was soon again triumphant. Admiral Pocock drove the French Commodore D'Ache out of the Indian seas, and Eyre Coote defeating at Wandewash the brave De Lally, who had commanded part of the Irish Brigade at Fontenoy, the power of the French in Hindustan fell for ever. In the West Indies Moro Castle was captured and Havana fell, and twelve ships of war rewarded the victors. The Windward Islands also became British. A fleet landed eight thousand soldiers at Lisbon, who drove the Spanish invaders over the frontier. Manila and the whole group of the Philippine Islands surrendered.

Here then, in mere outline, are abundant examples of how her sea power enabled England—a small Power as far as her insular resources were concerned—to strike blows in every part of the world, and to act in the short period of five years on the most distant points without hesitation or without any loss of communications. It is desirable to repeat that there never was any more striking display of military geography as affecting war in the history of the world.

On the other hand, the value of individual bravery was exemplified by our adversaries. Privateers in these years did as much damage to our commercial marine as did the Alabama to Federal shipping during the civil war. Though the French had not one naval squadron afloat in 1761, their privateers took no less than 812 English vessels.

Two more periods of our history must suffice to illustrate the marvellous manner in which naval power has enabled our islands to hold their own and to preserve and enlarge their possessions against any combination of enemies, even when these found allies among our own people.

Let us take the position of affairs in the period from 1778 to 1782. The American colonists declared their independence in 1776, and by the year 1779 France and Spain had also taken up arms against us as their allies. In Ireland the so-called Volunteers still further hampered our policy. Hyder Ali from Mysore threatened us in India, as did also the Mahrattas. The armed neutrality of which the leading members were the Dutch was really a confederation against our "Right of Search." Our nation, which then had only 12,000,000 inhabitants and a revenue of £9,000,000, fought in Virginia and Carolina, at St Eustatia and near St Lucia in the West Indies, at Pollilore near Madras, and in the heart of India against the Mahrattas. One of our generals lost Minorca and another (Elliot) saved Gibraltar. Admiral Hughes off Trincomalee and Cuddalore checked the able efforts of De Suffren to rehabilitate his nation's influence in the Indian seas. We lost the thirteen American colonies in 1781, when Cornwallis was besieged in Yorktown by Washington and Lafayette. Admiral Graves allowed Admiral de Grasse to gain a very temporary command of the sea on the Virginian coast. As Washington very wisely remarked in a letter to de Grasse, "The general naval superiority of the English previous to your arrival gave decisive advantages in the South in the rapid transport of their troops and supplies, while the immense land-marches of our succours, too tardy and expensive in every point of view, subjected us to be beaten in detail....Whatever efforts are made by the land forces, the Navy must have the casting vote in the present contest[1]." After this it is amusing to see

[1] Mahan, *Influence of Sea Power on History*, pp. 392—400.

the medal with the infant American Hercules strangling
the British serpent. The serpent would have crushed
the infant in its cradle but for the prompt and able
assistance of its French nurse.

In the closing years of the last, and the first year of
this century, the British displayed unparalleled activity in
every sea. As usual, there were many expeditions to
the West Indian Islands between 1796 and 1807. It
will be an evil day for the future of our empire and
our trade when, for whatever reason, we lose touch with
the ports and fertile lands which stud the Caribbean Sea.
We also despatched expeditions to Buenos Ayres, the
Cape of Good Hope, Egypt, Sicily, Calabria, the Dardanelles, and Copenhagen. In India we fought Tippoo
Sahib at Seringapatam, and the powerful Mahratta chiefs
at Assaye, Argaum, Delhi and elsewhere. An insurrection in Ireland was suppressed. Decisive naval victories
were won in 1797 over the Dutch at Camperdown and
the Spanish at St Vincent, and over the French at the
Nile in 1798, while the gate of Syria at Acre was closed
to Bonaparte in the following year. We occupied
Malta, and the new colony of Tasmania was founded in
the southern seas. We won the strategic victory of
Finisterre and the decisive battle of Trafalgar. These
marvellous efforts of our nation all over the world
when its population was only 16,000,000 excited the
amazement and admiration of Bonaparte, and as early
as 1797, at the commencement of his astonishing career,
he wrote,—" Either our government must destroy the
English monarchy, or must expect itself to be destroyed
by the corruption and intrigue of those active islanders.
Let us concentrate all our activity upon the navy and

destroy England. That done, Europe is at our feet."
Europe was at the feet of Napoleon early in 1812, but
the relations of the active islanders with the rest of the
world were as widespread as when the renowned
emperor had been plain Citizen Bonaparte and an
agent of the Directory. Mauritius had been annexed,
Java had been captured, we entered into a land and sea
war with the United States, we fought in Canada, we
occupied Sicily, we blockaded the ports of France and
her allies, we moved our soldiers into the centre of the
fastnesses of Nepaul; Wellington captured Ciudad
Rodrigo and Badajoz, and our soldiers and sailors were
at Lissa and Corfu. British commerce refused to die
under the far-reaching and desperate policy of the
continental system; on the contrary, it throve, to the
ruin of all competitors. And, notwithstanding all these
long-continued wars and rumours of wars, the state of
the people at home was so much more happy than the
lot of dwellers in any other lands, that Burke exclaimed,—"Our houses are bursting with opulence into
our streets." An eminent foreign writer felt bound to
confess that "there does not exist and never has elsewhere, so beautiful and perfect a model of public and
private prosperity."

Thus were justified after three hundred years the
practical advice of Raleigh and the deep philosophy of
Bacon. The melancholy ocean, which in the time of
the Romans, separated the Britons from the rest of the
world, had become a high road, with branches entering
into every bay and river-mouth, from Cathay to Vancouver Island, and from Lisbon to Quebec. Seas and
rivers alike were turned into avenues for British trade,

and from their banks the challenges of red-coated sentries warned all intruders that they were carefully guarded by the armed retainers of a world-wide empire. Yet in those days, the days of the Georges, there were no steamboats or railways.

These facts explain our present position and the marvel that 40,000,000 of Britons rule 400,000,000 of other folk, that an area of 121,000 square miles holds sway over 11,000,000 square miles, that our capital at home amounts at the most moderate estimate to £12,000,000,000, and that our language is spoken by 100,000,000 people. Even if luxury or political cowardice were to bring about the decay of the United Kingdom itself, nothing can prevent the Anglo-Saxon race, with its vigorous offshoots in Australia, North America, and South Africa, from domination over a third of the globe[1].

During the reign of Queen Victoria no fewer than 10,000,000 emigrants have left our shores to aid in the foundation of the empires of the future. In striking contrast to emigrants from other lands, these colonists in British possessions get nothing but naval guardianship from the people at home; while on the other hand, they contribute enormous sums to the private wealth of the mother country. They have asked for the very minimum of aid to start their enterprises. In most cases when they were dwelling in the wilderness they sought no help, and were offered none, and their mother country, jealous of the liberty of her children, has been as eager to grant as they have been to establish practical autonomy. But if all these connections with distant lands were desirable in the last century, when our people

[1] Niox, *Expansion Européenne.*

could live on the produce of our soil, they are vital matters now, when the greater part of our necessaries comes from abroad.   It was then expansion and wealth; it is now expansion or death.   Moreover, since the intro-duction of the use of steam, our manufacturing popula-tion has grown to such an extent, compared with our own agricultural population—and compared with the state of centres of population in other lands, ours being the only country where the urban folk are more numerous than the rural—that, unless we can keep open all existing markets, our workers cannot find employ-ment, and unless we can find new markets, the next generation must not marry or their children will perish. It is thus almost criminal folly to neglect the teach-ing of political, commercial, and military geography. Foreigners see clearly how we stand, and teach their students the details of our colonies, the sources of our wealth, and how vulnerable we have become, as thoroughly as Elizabethan writers discussed the state of the Spanish empire.

One French writer has recently undertaken to prove from geography that our empire and commerce will soon be crushed out of existence between the soldiers of the Old World and the navies of the New.   Such a future is by no means impossible.   An alliance between Powers like Germany or Russia, and the United States against us might well bring about as critical situations as existed in the periods 1797—1799 and 1803—1805. However this may be, the words of the eminent geo-grapher G. Niox should be well weighed.   He con-tends that apart from all questions of ambition, glory, or international status, the colonial expansion of this country is the necessary consequence of its industrial position, inasmuch as it produces yearly far more goods than its own people can consume. It also produces much less food than its operatives require, even if they were moderate and temperate in their requirements, whereas in point of fact, they are far from being either, and the servants and middle classes, with regard to diet, are relatively the most luxurious of mankind. With its square miles of mills and its myriads of workers, it is condemned to perpetual production if its fixed capital is not to rust out, and its circulating capital to lie stagnant, the families of the labouring classes to starve, and those of the mercantile classes to be reduced to a diet of pulse.

The existence of the operatives depends on our lines of communication by sea being left open and on "open doors" for our merchandise in every clime. Englishmen, therefore, have before them the duty of always concerning themselves with the opening of new markets and the search for new customers—of making, in short, their industrial machinery as complete and perfect as possible, so as to strive with success against their foreign competitors in Asia and Africa. They have not so concerned themselves for a generation; and with all their strategic advantages by sea, they will be ousted unless the new generation can sell mind as well as matter, and cultivate brains as well as muscles.

Shakespeare speaks of England as being a jewel set in a silver sea, but Niox's description of the country is now more exact:—"It is a block of iron and coal as isolated as was Ultima Thule." But the wisdom of our ancients planted sources of demand for the products of iron and coal in combination in many a colony and subject realm. The products of our looms and anvils flow in a perpetual stream into our ships, which make up half of all the trading vessels of the world, and bring back food and numberless articles of luxury from more genial climes.

These considerations now form the key to the solution of our foreign policy. We must see that the £1,100,000,000 worth of goods annually carried in the 36,300 vessels of our mercantile marine be not delayed *in transitu* by any hostile force; we must support a very powerful war fleet, and maintain coaling-stations, re-victualling-stations, and repairing-arsenals along all the great ocean lines of communication.

An interesting fact in connection with our empire is that the population under our rule in Africa already equals in number that of our Indian subjects in 1801, and is increasing in numbers and wealth at a very rapid rate. The present condition of the British empire, on the authority of Sir R. Giffen, may be summarised as follows :—

"The Empire, as thus viewed, is a territory of 11,500,000 square miles, or 13,000,000 if we include Egypt and the Soudan, and in this territory there is a population of about 407,000,000—which would be increased to over 420,000,000 if Egypt and the Soudan were included—a population about one-fourth of the whole population of the Earth. Of this population again, about 50,000,000 are of English speech and race, the ruling race, in the United Kingdom, in British North America, and in Australasia ; and the remaining 350,000,000 to 370,000,000 are the various subject races, for the most part in India and Africa, the proportion of the governing to the subject races being thus about one-eighth.   South Africa is an exception, being self-governing, with a white minority in power, but with the black subjects greatly predominating in numbers.

"The increase in area and population in this Empire, again excluding Egypt and the Soudan, amounts, since 1871, to 2,854,000 square miles of area, or more than one-fourth of the whole, and to 125,000,000 of population, which is also more than one-fourth of the whole.   The increase of the ruling race included in this population amounts to about 12,500,000, or about one-fourth of the number in 1897; and the increase in the subject races is 112,000,000, or nearly one-third the numbers in 1897. The increase in these subject races is largely, but by no means exclusively, due to annexation.

"The present revenue of the different parts of this Empire added together amounts to £257,653,000, and the imports and exports to £1,375,000,000.   The increase since 1871 also amounts to £115,143,000 for revenue, or more than 40 per cent. of the present total, while the increase in imports and exports amounts to £428,000,000, or about one-third of the present total.   The latter increase is perhaps greater in appearance than it really is, as all the figures are not reduced to a gold valuation, those for India for instance being in tens of rupees; but it has also to be considered that the gold valuation itself, owing to the increase in the purchasing power of gold since 1871, prevents the real growth of almost any economic factor being fairly shown by values only.   The Import and Export figures are also subject to the ob-servation that the trade of each part of the Empire is largely with other parts of the Empire, so that for some purposes they ought not to be added together. The revenue of the self-governing English portions of the Empire also amounts to £145,000,000, having increased £60,000,000 since 1871, and the imports and exports of the same portions to £1,036,000,000, having increased £247,000,000 since 1871. The revenue of the states of subject races also amounts to £112,000,000, having increased £55,000,000 since 1871, and the imports and exports to £338,000,000, having increased £181,000,000 since 1871[1]."

Another very important feature of our empire, in which it differs from the Roman and Spanish empires at the height of their power, is that the numbers and resources of the governing race increase at a more rapid rate than those of most of the subject races. This, unless the ruling race decays physically and morally, is a security for the permanence of its power.

With regard to France, the position since the beginning of the century has changed enormously to our advantage. But we cannot ignore the dangers which threaten our own future position from the phenomenal expansion in territory, population, and other resources of the United States and Russia, and the military power of the United German Empire, the high educational standard of its people, and the new-born maritime and commercial activity which now makes its towns worthy successors to the old Hanseatic league.

From a strategic point of view, Ireland is of the utmost importance to Great Britain, and must hence never be allowed to fall into foreign or hostile hands.

[1] *Journal Royal Colonial Institute*, March, 1899.

It is a land-barrier of 300 miles midway between the Pentland Firth and the English Channel, it commands all the Atlantic approaches to Great Britain, and our navy in the Irish Sea can harass hostile forces approaching by the northern or southern entrances alternately, or at the same time.

That intimate relations, owing to our unprecedented sea power, can be established between different parts of our empire was proved by the fact that Canadian voyageurs managed the boats on the Nile during Lord Wolseley's advance towards Khartum, and an Australian contingent arrived at Suakin in 1885.

The powerful influence of British naval supremacy in regard to international complications was admitted by Germany when it displayed feverish anxiety that we should join the Triple Alliance.   General Maurice[1] says that the adhesion of our fleet would have been at least as valuable to the Alliance in case of war as reinforcements to the amount of 300,000 men.   American authorities are unanimous in admitting that, whatever the motive might have been, our Government in 1898 conferred priceless benefits on their country.   France and Germany became strong adherents of Spain, their newspapers wrote incessantly against the politics of the people of the United States.   France may have been influenced by the fact that much of its capital was invested in the Spanish debt, and the United States' tariff engendered much animosity among the German cultivators.   The Austrian Monarchy, for family and historic reasons, would also naturally favour Spain, " but Great Britain from the start took strong sides

[1] *Balance of Military Power*, 1888, p. 195.

with the United States, evincing an unexpected warmth of friendliness and a strong desire to ally itself with this country. Whatever the underlying motive in the British heart, this earnest display of friendliness was of much service to the United States. It tied the hands of our enemies on the Continent, who feared that any hostile act would result in an alliance between the two great Anglo-Saxon nations. Some active efforts at interference might have been made but for this haunting fear—Great Britain stood as a buffer between us and our opponents; she refusing to co-operate in any steps of interference[1]." In a similar manner, Great Britain in 1862, by refusing to join the Emperor Napoleon III. in his schemes against the United States, did much to prevent their being ruined by the Secession States.

Canada has waited a good many years for equal opportunities with the United States in rapid and frequent communication with this country, and she is now in a fair way of winning them. In the past both her produce-trade and her passenger traffic were handicapped by the frequent necessity of maintaining the connection with home through New York, which to all intents was a foreign and almost a hostile port. For many years the sole line of passenger steamers that sailed out of Liverpool for Canadian ports was the Beaver Line, which was not to be compared for speed and sumptuousness with the lines running to New York. At that period neither Canadian produce nor Canadian passengers bulked largely in the estimation of Liverpool shipowners. Now, all this is changed. Canadian perishable produce demands as quick transit as that of the

---

[1] Morris, *War with Spain*, Chap. VII. p. 149.

United States. The Canadian Steamship Line was started in December 1898, to run from Milford Haven —with fast trains from Paddington—to Paspebiac, a new port at the mouth of the St Lawrence on the north of Prince Edward's Island, and its vessels are expected to perform the voyage in from four and a-half to five days. Another fast line of steamers, the vessels of which, it is said, will be the finest ever seen in a Canadian port, is about to start from Liverpool with a weekly passenger service to Montreal. The more of these lines the better both for Canada and the home country, for it will assuredly be found that greater facilities for trade and intercourse will increase both, and in case of war, the value of lines of communication open all the year from Liverpool to Esquimault would be incalculable.

Indifference to colonial expansion at one period of our history brought us nearly to the point of allowing British Columbia as well as Alaska, Washington, and Oregon to be annexed to the United States. This would have been a serious loss to Canada and to our empire at large, but in 1871, by the energy of Sir John MacDonald, this splendid province was definitely secured and the trans-continental railway extended to the coast. The result has been of the first strategic importance, independent of the auriferous and other natural wealth of the country. It gives us fine ports on the Pacific coasts. Moreover it projects towards Asia, and the route from Liverpool to Asia by Vancouver is 6,000 miles shorter than that by New York. To quote the *United Service Magazine:*—"...these advantages, great as they are, would be of little value without good coal, of which British Columbia possesses an inexhaustible supply, whereas on the coast it is poor in quality and limited in quantity. Esquimault, fortified by Canada and England, is thus a naval base of the utmost importance, in fact the key to Imperial supremacy in the North Pacific. Eastwards there is no other coaling-station nearer than Hong-Kong, southwards, none nearer than Fiji. In time of war it will serve the English navy to as good purpose as the want of it will completely paralyse the fleet of an enemy. Opposite stands Vancouver, the terminus of the Canadian Pacific Railway, by means of which England can land troops in China in half the time it would take by way of Suez or the Cape[1]."

Some idea of the Dominion's spirit and confidence in her destiny may be gained from the fact that her greatest iron road was built, while its terminus was still an uncertainty, across a vast wilderness over a thousand miles long, and at an expenditure of £24,000,000. Surely never before did three millions of people undertake a public work so stupendous or so little likely to be pro-ductive except in the remote future. That they were colonials reflects fresh glory on the achievement, and is one of those stubborn facts which a cynical world can never explain away. Moreover nearly all the great public works of Canada have been conceived and carried out in harmony with Imperial interests. Therefore they are as necessary to the defence of the British Empire as they are to the defence of the Dominion herself.

The distance between Calcutta and Esquimault is 9100 geographical miles. For average trooping purposes 13 knots an hour may be allowed, and it may therefore be estimated as a 29 days' journey; but the speed might

[1] *United Service Magazine*, June, 1898.

be 15 knots, or 360 miles a day, in which case 26 days
would suffice. The Canadian Pacific Railway journey
would take six days at least, including embarkation and
disembarkation. From Halifax to Liverpool is 2490
miles, which at 15 knots would take seven days. To
transfer troops from Liverpool to Calcutta by Canada
would therefore take 39 days.

From England to Calcutta by the Cape is an affair
of 35 or 36 days. The mail to Bombay takes 14 days,
but here we have to do with an overland route and very
fast steamers, and the Suez Canal. Perchance the
railway to Salonika may still further accelerate the
transit.

Outside the British Isles proper, we have *points
d'appui* in the following places :—The Channel Islands.
In the Atlantic Ocean:—St John's, Newfoundland; Halifax, Nova Scotia ; Hamilton in the Bermudas, Nassau
in the Bahamas, Kingston in Jamaica, Antigua, Bridge-
town in Barbados, Port of Spain in Trinidad ; George-
town, Demerara. On the line to India by the Suez
Canal:—Gibraltar, Malta, Cyprus, Egypt, Aden, Bombay,
Colombo, Madras, Calcutta. On the road to China :—
Colombo, Penang, Singapore, Hong-Kong, also Labuan,
and Kuching in Sarawak. On the road to Australia :—
Aden, Mahé (Seychelles) Mauritius, Adelaide, Melbourne, and the great depôt of Sydney. On the West
and East African lines :—Bathurst (Gambia), Freetown
(Sierra Leone), Ascension, St Helena, Simon's Bay, Port
Elizabeth, Durban (pre-emption rights as to Delagoa
Bay), Zanzibar. Besides the West Indian ports, we have
also in the New World :—Port Stanley in the Falkland
Islands, and Esquimault in Vancouver.

If it be true that our people spend a large amount on the Navy, though the sum is small relatively to that spent by other Powers, it must be admitted that the Navy, by protecting such vast areas of traffic, justifies its existence, and gives us a very good return for our money.

CHAPTER VII.

THE STRATEGIC RELATIONS OF EUROPE WITH OTHER CONTINENTS.

TAKING Europe as a whole, the northern frontier from Norway to the White Sea may be regarded as absolutely secure; and its western parts, protected by the Atlantic, were till quite recently considered perfectly safe as against any other continent. Although the course of Empire had frequently taken its way westward, though the New World had been called into existence to redress the balance of the Old, though Spain had by the instrumentality of Pizarro and Cortes annihilated the empires of Montezuma and the Incas, and though our own fleets had transported in the old sailing days large expeditions to Boston, New York, and Charleston, and carried soldiers up the St Lawrence to Quebec, it never had appeared at all probable that Europe might yet have no small difficulty in holding its own against the United States of America until 1898.

But now several European Powers besides Spain would be glad of an opportunity of co-operating against the United States if a favourable opportunity presented itself, and it is an open secret that, quite independent of any questions that might arise if an Anglo-American alliance were concluded, a European continental alliance against the Western Republic, based on agrarian and tariff problems, has frequently been advocated in Berlin and Vienna. The questions of mere distance, however enormous they may appear to stay-at-home students, have never frightened a man of genius. Mongolia to Moscow is as far as from the United States to Spain. From Southampton to Bombay by the Cape of Good Hope is four times the distance of from Cork to Halifax. Spain conquered Mexico; England conquered India. From France to Tonquin by the Suez Canal is twice the distance of from Brest to Charleston. The geographical conditions which would affect a transatlantic invasion of any part of Europe were well considered soon after the close of the American "War of Secession" by the Italian geographer Sironi. He held it by no means impossible that a serious menace to Europe might come from America, having regard to the prodigiously rapid development of the life and power of the great Republic. Since he wrote, in numbers, wealth, and potentialities of future progress, its power has enormously increased, and it has already crushed one European State. Moreover the ablest naval author in the world, Captain Mahan, in May 1897, wrote an essay urging on his countrymen the necessity to share in the trophies of the "general outward impulse of all the civilized nations in the first order of greatness." He proved the necessity of seizing upon some port between Vancouver and Australia.

"The serious menace to our Pacific coast and our Pacific trade, if so important a position as Hawaii were held by a possible enemy, has been mentioned frequently in the press, and dwelt upon in the diplomatic papers which from time to time are given to the public. It may be assumed that it is generally acknowledged. Upon one particular, however, too much stress cannot be laid, one to which naval officers cannot but be more sensitive than the general public, and that is the immense disadvantage to us of any maritime enemy having a coaling-station well within twenty-five hundred miles, as this is, of every point of our coast-line from Puget Sound to Mexico. Were there many other available, we might find it difficult to exclude from all. There is, however, but the one. Shut out from the Sandwich Islands as a coal base, an enemy is thrown back for supplies of fuel to distances of thirty-five hundred, or four thousand miles,—or between seven thousand and eight thousand, going and coming,—an impediment to sustained maritime operations well-nigh prohibitive. The coal-mines of British Columbia constitute, of course, a qualification to this statement; but upon them, if need arose, we might hope at least to impose some trammels by action from the land side. It is rarely that so important a factor in the attack or defence of the coast-line—of a sea-frontier—is concentrated in a single position ; and the circumstance renders doubly imperative upon us to secure it if we righteously can[1]." Hawaii has since been annexed by the United States.

Mahan moreover deplored the fact that the finest West Indian island was in hands that were not equal to its requirements, inasmuch as the cluster of island fortresses of the Caribbean Sea is one of the greatest of the nerve-centres of European—indeed, of cosmopolitan—naval strategy.

[1] Mahan, *Interest of America in Sea Power*, p. 47.

"The intrinsic advantages of Cuba are pre-eminent, and also, but in much less degree, those of Great Britain in Jamaica. Cuba, though narrow throughout, is over six hundred miles long from Cape San Antonio to Cape Maysi. It is, in short, not so much an island as a continent, susceptible, under proper development, of great resources—of self-sufficingness. In area it is half as large as Ireland, but, owing to its peculiar form, is much more than twice as long. Marine distances, therefore, are drawn out to an extreme degree. Its many natural harbours concentrate themselves, to a military examination, into three principal groups, whose representatives are, in the west, Havana; in the east, Santiago; while near midway of the southern shore lies Cienfuegos. The shortest water distances separating any two of these is three hundred and fifty-five miles, from Santiago to Cienfuegos. To get from Cienfuegos to Havana, four hundred and fifty miles of water must be traversed and the western point of the island doubled; yet the two ports are distant by land only a little more than a hundred miles of fairly easy country[1]." Since these words were written Cuba has ceased to belong to Spain. Philanthropy at the end of this century is as fertile in annexations as was ambition in past ages.

While the expansion of America westward and eastward has thus altered the future of international strategy and of political geography, the extension of Russia eastward has obliterated the eastern frontier of Europe from every practical point of view. The Caucasus, Ural Mountains, Ural river and Caspian Sea are mere local geographical terms of no present strategical importance.

[1] Mahan, *Interest of America in Sea Power*, p. 289.

Russian military power is moving across Siberia by rail
to the Pacific, nor will the ice any longer hamper her development; a railway will ere long join the Amur to Port
Arthur.   There is a Russian railway from the Black Sea
to the Caspian, and thence it is extending to the frontiers
of India.   Cossacks have replaced the Golden Horde.
The people of Europe have avenged their ancestors ;
Muscovites hold sway in the districts whence, as fiercely
as from Tartarus itself, Moslem heroes issued to the
horror of the West.   Russian pickets even now post
sentries among the Pamirs, on the " Roof of the world,"
and the soldiers of the Czar guard the tomb of Timurlane in Samarkand.   The old European boundary
therefore has ceased to be of the least strategic significance, and the defiles of Ekaterinburg, the Ufa, Orenburg,
and Dariel, and the roads from Persia to Asia Minor
and Syria, are now of purely archæological or academic
interest, and unless China can be resuscitated and become
a strong man armed, must so remain.   As Napoleon
said when justifying his raid into Egypt and Syria,
" Europe is exhausted"; he challenged the right of
Britons and Russians to omnipotence in the East.
He failed, but since his time United Italy and United
Germany have been added as competitors to the British,
French, and Russians for the choicest regions of Asia and
Africa.

      With regard to the southern frontiers of Europe the
time was when Numidians and Carthaginians and Moors
more than held their own with the rulers of the northern
and peninsular coasts of the Mediterranean.   But every
strategic position on the African shore, from the " brook
that parts Egypt from Syrian ground " to the pillars of

Hercules, is now dominated by European force, and if any fortresses be built and any military schemes concerted in the land once defended by Hasdrubal or Jugurtha or the Ptolemies, it will be by Europeans against Europeans, and not against Africans.

It is therefore more practical to begin our studies with the configuration of the western European States and thence to work our way inward and eastward by the Baltic and the main routes of Central Europe and the Mediterranean. The western front of Europe is represented by a line joining the extreme north of Scandinavia to Cape St Vincent in Portugal, in length 2700 miles—about the same distance as from Gibraltar to Tiflis, or from Crete to Novaya Zemlya, or across the Canadian Dominion. Along this measurement three salients are pushed into the ocean—the Scandinavian coast, the French coast, and the Spanish-Portuguese coast. These form three large bastions, while the British Isles are a kind of advanced work covering all the lines of invasion into northern Europe by the Channel, the North Sea, the Baltic Sea, and the river-basins opening into these seas. Scandinavia forms a third of the whole distance, but its coasts and mountain ranges are quite impracticable, and except as concerns its southern shores and their relations to the entrances of the Baltic it has no strategic interest with regard to its own interior, or to avenues leading to less mountainous and more prosperous districts. From the fjords of Scandinavia daring adventurers have, in past times, issued to ravage all the coasts of Europe, and sometimes to found inland States as in England and France and Russia, or, as in the days of Gustavus

Adolphus and Charles XII., to throw Swedish swords
into the scales of international controversy, but the
conquest of Sweden and Norway has presented few
temptations to more southern nations.

Any invader who passes the Straits of Gibraltar can
penetrate to the heart of Central Europe from the Gulf
of Lyons or the Adriatic, but the Iberian peninsula
commands the passage, and flanks its entrance and its
exit. Therefore Spain could play a leading part either
in the attack or the defence of Europe at large. The
master of Spain now commands the western Mediterranean and also the basins of the Garonne and the
Rhône, as at the close of the Peninsular War and as in
the days of Hannibal. From Spain the Moors threatened
Christendom and advanced as far as the Seine, the Jura,
and the Alps. Moreover the Peninsula has a most
advantageous situation with regard to the New World
and the seas which wash the coasts of Africa, Persia,
and India. The discovery of the West Indian Isles and
of the passage round the Cape of Good Hope, and the
founding of the first European factories on the African
coasts, in Hindustan, and in the Asiatic archipelagoes
prove how far strategic geography was once the handmaid of Iberian enterprise.

But neither Spain nor Portugal is likely to be a
dangerous enemy for some time to come, while the
enormous difficulties in the way of an invader—difficulties which render this territory most instructive to
the student of strategy—will prevent Spain being used
with any approach to promptitude as a base of operations
against any other European Power. Of course if the
" Pyrenees ceased to exist " the harbours of Spain would

be of the utmost value to France, and the union of the two States would produce a dangerous situation for the world at large, a situation which was only prevented by the constant efforts of Great Britain and Austria, and Portugal and Piedmont, from 1696 to 1713. It will be seen that in all political disputes in which Spain is concerned the question of sea power is of the greatest consequence. If an invader once gets hold of Spain it would be hard to eject him. The Moors bade defiance to all the courage of a succession of heroes for seven centuries, and when their power was at last broken they stood at bay for a long time in the valley of the Guadalquivir and in the fastnesses of the Sierra Nevada before making up their minds to recross the straits for their old African homes.

The middle section, then, of the Western front of Europe would clearly, as far as natural obstacles are concerned, be the most convenient for the invader. The line from the coasts between the Bay of Biscay and the Baltic is the simplest, but this line is re-entrant and flanked by two peninsulas, and is protected by the British Isles. The occupation of these isles and the northern section of Spain would be an indispensable preliminary to a grand invasion of central Europe from the Western Ocean. The occupation of France would greatly facilitate the invader's designs. A strong French army based upon this fertile soil, even after the wasteful excesses of Revolution, could and did work marvels.

Between 1805 and 1809 every Power in Europe, from the Channel to the Ural Mountains and from the Baltic to the Mediterranean, became either the subject or the obedient ally of Napoleon, with the exception of Portugal

and Spain, and their struggle would have been futile
were it not that the wealth and the military and naval
resources of the British Isles were placed at their disposal.   A glance at the map of Europe will suffice for
the present to set forth how the possession of the basins
of the Seine, the Loire, and the Rhône would give
strategic *points d'appui* against North-Western Europe,
Switzerland, and Italy.

The Baltic Sea can easily be closed to the invader;
the Danish straits can be blocked at will; indeed, although some 50,000 merchant vessels annually pass the
Sound, modern ships-of-war of the first class could not
do so, and would be confined to the Great Belt and soon
involved in difficult navigation caused by the currents
and the labyrinth of little islands.   These straits are
from seven to twelve miles wide.   The British had no
difficulty in the Napoleonic wars in enforcing their will
on the Baltic Powers.   They crushed the Armed Neutrality in 1801.   After Nelson's great victory at Copenhagen over Denmark, Russia gave in, and agreed to the
British right of search and to the capture of enemies'
goods in neutral vessels, and gave up the doctrine of
paper blockade.   During the period from November
to April the Baltic seaports are frozen, and another
difficulty in the way of the invader is the peculiar
character of the North German coast.   It is too sandy
and marshy, and its entrances from the sea too full of
bars and other obstacles, to render it possible that
ships of the first or second class could operate successfully either by bombarding such fortresses as exist at
Königsberg and Dantzic and the mouth of the Oder,
or by disembarking considerable expeditionary forces.

Though the British under Nelson were able to terrify the court of St Petersburg in 1801, they could not do so in 1854. The allies had some successes at Bomarsund, but their attack at Cronstadt was a decided failure. It might have been that the gunboats which were so lavishly and hastily provided after this failure for the special circumstances of the Russian coast would have justified their existence had the war been prolonged, but the ignoble Peace of Paris in 1856 closed the war by conceding, for no sound reasons, the very points for which we fought the Armed Neutralities of 1782 and 1801. Another failure in the Baltic was that of the French fleet in 1870. Admiral Bonet-Willaumez was ordered to detach a force to watch the very small German fleet at the mouth of the Elbe, and to proceed to harass the Baltic coasts and thus divert some corps northward from their march to France. One division and the local reserves were quite sufficient, and the French returned having effected nothing. Even if the Admiral had the troops necessary he had only large ships of war and no small vessels, though the experiences of fourteen years before might have given some lessons to the French Admiralty. He was brought home, and his sailors were more usefully employed in working guns in the fortresses, which the German invaders promptly besieged.

The German Ocean is much easier for fleets, and accordingly the Germans have linked the Baltic from their great arsenal at Kiel to the mouth of the Elbe by a ship-canal, but though the German Ocean is not a British *mare clausum* as Selden and Cromwell contended, it is under British command, owing to the enormous trade done by the towns on our Eastern coast and to the fact that no hostile fleet can pass the narrow straits of Dover without a fight. There can be no possibility of eluding our fleets between Plymouth and Harwich. Since 1588 no fleet has been able to do so. The Dutch fought strenuously for twenty years to secure freedom of access from the Zuyder Zee and Rotterdam to their Asiatic dependencies, but in vain. In spite of some of the very hardest naval battles in history in the waters between the Cinque Ports and the Wash they failed utterly in the days of Cromwell and Charles II, and from 1689 till 1789 the Dutch were under British influence and owed what was left of their greatness to the British sword. Thus it would be impossible, so long as the United Kingdom is a powerful Naval State, for any nation not in alliance with her, or which had not secured her neutrality, to produce any effect on the fortunes of Northern Europe by sea power.

The Mediterranean is also a closed sea, as is the Sea of Marmora and the Black Sea. While Admiral Hornby could stop the Russians at San Stephano in 1878 they were starving, because their ships dared not approach the coast between the mouths of the Danube and the Bosporus. Neglect of their navy by the Turks since 1878 has been their greatest strategic blunder. A British fleet at Besika Bay would command the Dardanelles. We hold Cyprus, which seems to Captain Mahan and the French geographer Niox the key of Egypt and Syria, though it is held in small esteem by many of our own authorities. But no one questions the value of the positions of Port Mahon which we once had, and of Malta and Gibraltar, now strongly fortified, and in

French opinion invulnerable, even if our fleets were obliged to leave them for a period.

It must be remembered that even before the construction of the Suez Canal, which has modified the whole current of international commerce as well as changed the centres of Sea Power, a temporary loss of our influence in the Mediterranean was highly detrimental to our interests. Still further caution is now required since the construction of works at Biserta by the French, which, but for Malta, would give them the command of the Sicilian strait between Sicily and Africa. The Italians, recognising the importance of this waterway, have fortified Maddalena and Elba and the Straits of Messina, but it would be dangerous to hostile fleets to venture past the fortress of Gibraltar if a British fleet were at hand, and impossible if the latter were allied with the Spanish at Cadiz. The French depôt at Toulon is extremely formidable, but it would be in the highest degree hazardous to try and join the Mediterranean fleet to the Atlantic fleet at Cherbourg, Brest, and the Gironde. Hence the scheme for a new maritime canal in the south of France from Bordeaux to Cette, as the existing canal can accommodate only torpedo boats. The Russian naval power in the Baltic is separated from the Black Sea fleet by a vast distance, and in war time Gibraltar and Malta could block the way to their junction.

To the new canal proposed to be constructed from the Gulf of Riga to Kherson on the Black Sea we have already alluded[1].

In less than ten years the Russian position in Manchuria will be unassailable, and Port Arthur, having regard to its geographical position, will be quite as strong as Cronstadt or Sebastopol. In substituting Port Arthur for Vladivostock as her principal naval station in the Far East, Russia gains immense advantages[1]. The latter has scanty local resources, and is practically an island dependent on an uncompleted single line of railway four thousand four hundred miles long for its communications, being thus far less favourably circumstanced than Hong-Kong, in the hands of a great naval and maritime power. The former possesses coal, iron, rich agricultural possibilities, and a hardy population capable of furnishing excellent military material. It is also easily defensible against naval attack; and with a railway and a well-organised army at its back, it will have nothing to fear from operations such as those of the Crimean campaign, undertaken by a European power at a vast distance from any home base. Finally, when the Manchurian railways are constructed, great military forces will be within striking distance of Peking. No territorial advance of Russia in the present century is comparable in importance to the step which has just been taken, after long and careful preparation.

The momentous results of the possession of sea power appear in the modern history of Russia[2]. The great stride in the Far East which has carried Russia from the banks of the Amur to the Gulf of Pe-chi-li was brought within the scope of practical politics by the creation of the fleet. The policy is precisely that of Peter the Great, only the method differs somewhat

[1] Sir G. Clarke, *Russia's Sea Power*.
[2] *Ibid.* p. 133.

with the changed circumstances of the times. It was
necessary for Peter from his inland state to conquer a
seaboard on the Baltic and the Euxine, and, when there
established, to build up a navy. Vladivostock, the
Russian station in the Far East, has for years been
supported from the sea ; and when the German descent
upon the province of Shantung was accomplished, and
the practical dismemberment of China began, it was the
possession of a powerful fleet which enabled the advisers
of the Tsar to antedate their plans by the occupation
of Ta-lien-wan and Port Arthur. In 1702–3, Peter's
troops overran Ingria and Livonia, and captured Note-
burg and Nyen, enabling a fleet to be built in the Gulf
of Finland. In 1898 the navy of Nicholas II, built in
Europe, established itself in the ports of the Liaotong
peninsula, and land forces are prepared to move to the
support of the navy.

CHAPTER VIII.

FRONTIERS IN THEIR RELATION TO MILITARY OPERATIONS.

At first sight it seems fortunate for a State when its boundaries are so determined by nature that it is secluded from other lands, and cannot be invaded except at great risks across stormy seas, wide and rapid rivers, or rugged mountains. A glance at the maps of Europe and Asia will suffice to illustrate the difference between natural and artificial frontiers. Hindustan is cut off from Central Asia by the Himalayas, the Pamirs, and the Hindu Kush Mountains. Its northern portion from Kashmir to Assam is manifestly closed to armies. To pass through Tibet from the north would be difficult; to follow up this enterprise by treading the few and narrow paths over which Everest, Dwalagiri, and Kinchinjunga are eternal sentries would appal the boldest of the lieutenants of Timurlane. The masses of mountains which extend from Beluchistan by Afghanistan and Kafiristan to the Karakoram Pass would also seem an area impossible for the march of large armies, and accordingly many writers ridicule the notion of an invasion of our Indian Empire from the north-west. The Pyrenees are a feeble barrier, whether their height, width or length, or the severity of

their climate be considered as compared with the northern
buttresses of India, yet they make an admirable natural
obstacle between France and Spain, which has kept these
two nations apart ever since the fall of that marvellous
military and geographical unit, the Roman Empire.
The supposed natural boundary of France in the east
and north-east has never—except for a brief space in
Napoleon's time—coincided with the actual frontier. Indeed, when the idea was propounded by the founders of
the Republic, one and indivisible, in 1795, it was treated
with the utmost ridicule by Burke in some of his ablest
disquisitions.   The Alps, the Jura, and the Rhine from
Basle downward to its mouth would form an excellent
frontier.   But, even before 1870, the Bavarian Palatinate
and Rhenish Prussia and Belgium excluded France from
two-thirds of the Rhine, and since 1870 no fragment of
French territory impinges on the fine river which in the
days of Julius Caesar marked the limits of the Gauls
and the Teutons.   In Italy the various Alps and the
Dinaric range, which would give Italy the Eastern coast
of the Adriatic north of Montenegro, suggest themselves
as natural boundaries, but in point of fact such a frontier,
though in the main continuous with the Alps from the
Cottian to the Julian, has many flaws, and is artificial in
several sections.   Race, language, and historical accidents
have fixed the boundaries of nations as much as geographical conditions.   When the limits of the Turkish
Empire, as was the case a century ago, extended from
the Adriatic to the Black Sea along the Save, down the
Danube from Eszek by Belgrade to Orsova, and thence
along the southern slopes of the Carpathians to the Pruth,
it had a fine strategic frontier covering a series of inner

entrenchments which secured Constantinople for ages.
Nor was the southern portion of European Turkey in
a worse case.   From the Morea to the Bosporus every
strategic issue or decisive point was in their hands, and
the hardy fanatics of the north-west of Asia, while the
Porte had command of the seas and straits of eastern
Europe, could swarm to the aid of their brethren, whether
fiercely resisting Christian chivalry on the Drave, or
driving back the Muscovites from Trajan's Wall to the
Dniester.   In 1854 these were foiled on the Danube,
and in 1877 they were well-nigh ruined between that
river and the Balkan Mountains.   The Danube between
its source and the Iron Gates at Orsova is not now the
boundary of any State.

The Romans extended their frontier over the Alps
so as to fix a barrier between their civilisation and the
dreaded barbarians of the North. The Emperors stopped
short, and in consequence they were ruined; they should
have pressed on Germany, shattered the Goths in their
own homes, and planted Roman laws and legions on
the Oder and Elbe, as well as by the banks of the
Moselle and the Thames. Nations must advance or
die.   The ruler who tries to fix a boundary to Empire
is generally engaged in a futile enterprise, and if perchance he succeeds, the result is quick decay.   A little
more perseverance under Tiberius and Hadrian would
probably have reduced all Germany[1], but they hearkened
to the fatal caution of Augustus, *concilium coercendi intra
terminos imperii.*

No mountain frontiers, therefore, have prevented
invasion or for long stopped the tide of war, but they

[1] Bryce, *Holy Roman Empire*, p. 12.

are of great strategic utility, inasmuch as, much more
than rivers, they limit the avenues by which the invader
can advance to his object, and they perplex him more
than does a river, because transverse roads from defile
to defile must always be less common than roads along
the banks of rivers from bridge to bridge. Moreover
they are valuable to screen the movements of a strategist
from his opponent, and thereby they favour surprise.
For example, in 1813, when Napoleon was coping with
the heroes of the German War of Liberation, he moved
eastward from the Elbe to the Oder, north of the mountains of Bohemia. But while he was only thinking of
danger from Blucher in his front, the Austrians moved
rapidly into Bohemia, and towards the passes leading
into Saxony, threatening his line of communications
with Dresden. He was thus obliged to counter-march
with rapidity, and was only just in time to save Dresden.

The dominion of the house of Austria would seem
to be enclosed with a mighty and impenetrable natural
frontier by means of the Transylvanian Alps and the
Carpathians with their Galician *glacis*, the mountains of
Bosnia and the Julian, Carnic, and Tyrolese Alps, the
Vorarlberg and the hilly curtain of the Salz and Inn
valleys, the Böhmer Wald, Erzgebirge, Riesengebirge
and Sudeten range. But the valley of the Danube and
its numerous affluents open up passages to Vienna from
the Rhine on the one side and the Balkan peninsula on
the other, while Prussian armies have had no difficulty in
traversing Bohemia and Moravia from secondary bases in
Saxony and Silesia. Many an invading force has poured
into the rich and prosperous valleys and plains of Hungary to revel in their wealth of wine and grain, horses and

oxen.   The roads from Cracow and Lemberg, Kronstadt
and Belgrade, are marked with the sites of many a battle
from the days of Attila to Hunniades, and from Prince
Eugene to Diebitsch many a national and religious controversy was referred to the die of war near Prague and
Königgrätz, and in the environs of Brünn, Olmütz,
Komorn, and Mohacs.   Napoleon used the road from
Ulm by Ratisbon, and from Augsburg by Munich in
1805 and 1809.   Even Bosnia is not safe, inasmuch as
the mountaineers of Montenegro are close at hand, and
being well used to warfare for centuries on rugged slopes
and on the sides of deep and gloomy ravines, could
seriously hamper Austria in a war with Russia.   Nature
therefore has here been assisted by a series of forts connected together by a new road practicable for artillery
and convoys.

   That men and not mountains determine the fate of
nations is further proved by the fact that long-continued
warfare has prevailed in most mountainous countries.
Our frequent wars north of the line from Kabul to
Attock and west of the Indus are examples.   More
instructive still were the campaigns of St Cyr and
Suchet in Catalonia.   The word " Pyrenees " is inscribed
on the colours of many a British regiment, but few
of our soldiers have any notion of the brilliant efforts
and desperate energy of both British and French in
the end of July and beginning of August, 1813, to
which the word refers.   Marshal Soult's efforts to force
Wellington's army from Pampeluna and San Sebastian
were very able, but he was driven back after a week's
fighting, in which he lost 15,000 men.

   Switzerland was the scene of very clever mountain

warfare in the days of the great de Rohan. In his campaign against the Imperialists in 1635 he was ordered
to conquer the Valtelline with 15,000 men. He began
operations by forcing the Duke of Lorraine, who had
passed the Rhine at Brisach, to evacuate Alsace. He
secretly crossed the Rhine near Basle, and after a march
of twelve days appeared before Coire, to the great joy
of the Grisons. At first his opponents had the better
of him, but, by a clever counter-march which brought
him to the heights of Cassiano, he surprised and routed
the enemy, seized the Valtelline, and retained it after
four battles against the generals of the Emperor. In
the following year he traversed and seized the three
valleys of the Milanese. Massena and Lecourbe conducted a striking series of operations against the
Austrians and Russians in the heart of the Swiss
mountains, between the Limmat and the upper Rhine,
in 1799. Massena having occupied the line of the Thur,
the Archduke Charles passed the Rhine at Stein and
another corps passed it at Feldkirch. Massena wished
to prevent their union but was obliged to retire to Zurich,
where he fought a great battle for three days. He then
moved to the left bank of the Limmat and occupied the
heights of Albis, his right on the Lake of Zurich, his left
near the confluence of the Aar, the Limmat, and the
Reuss. He remained three months on the defensive,
while Lecourbe executed a variety of movements which
are a model of mountain warfare. He nevertheless was
obliged to leave the Engadine and fall back to the
St Gothard. In September Massena took the offensive,
repassed the Limmat, and drove back the Austrians.
Meanwhile Suwarrow forced the St Gothard from Italy

so as to join the Austrians. He arrived at Altdorf, but his allies were beaten and he was boldly faced by Lecourbe, who thus forced him to pass the Alps of Glarus over a by-path encumbered with snow. After the loss of most of his guns and baggage he reached Germany in safety with the remnant of his army.

By mountain warfare is not meant the mere attack or defence of a mountain pass, such as we read of in the Tyrolean insurrection of 1809; but the attack and defence of a whole mountain country, comprehending perhaps a line of eighty or a hundred miles. Here almost all the elements of interest of war are combined; the highest exercise of skill in the general in the planning of his operations; the greatest forethought and energy in the officers and soldiers in overcoming or turning to account the natural difficulties of the ground. In such warfare, a general must bear constantly in mind the whole anatomy of the mountains which he is defending or attacking; the geographical distance of the several valleys and passes from each other, their facilities of lateral com- munication, their exact bearings and windings, as well as the details of their natural features and resources. Wellington ascertained many of these points by personal observation, spending hours in the saddle to this end. A general must also conceive the disposition of his enemy's army, its strength at each particular point, and the facilities of massing a large force at any one point in a given time. Bonaparte studied all the operations of Berwick and Maillebois before venturing into the Apennines. For a blow struck with effect at any one spot is felt along the whole line, and the strongest positions are sometimes necessarily abandoned without firing a shot, merely because a point has been carried at the distance of thirty or forty miles from them, by which the enemy may penetrate within their limits or threaten their rear[1]. And surely the moving forty, fifty, or seventy thousand men with such precision that, marching from many different quarters they may all be brought together at a given hour on a given spot, as was the case at the Zadora in 1813, is a very magnificent combination if we consider how many points must be embraced at once in the mind in order to its conception, and how many more are essential to its successful execution[2].

The celebrated "*Ne Plus Ultra* lines" stretched from Namur on the Sambre and Meuse to the coast of Picardy. The object was to keep the Allied Forces beyond the interior lines of fortresses which covered the frontier on the side of Arras and Cambray. From the left they ran along the marshy banks of the Canche, supported by the posts of Montreuil, Hesdin, and Frévent, and in front were the fortresses of Dunkirk, Gravelines, Calais, and St Omer. The Canche was connected with the Gy by redans; the Gy and Scarpe were checked by dams causing inundations. A canal of communication was opened from the Scarpe to the marshes of the Sanzet near L'Ecluse; there were forts at Aubigny, Pallue, and Aubanchoil; a fortress at Bouchain and a *tête-du-pont* at Denain. The course of the Scheldt was thus covered to Valenciennes, while further entrenchments to the Sambre were supported by Quesnoy and Landrecies. Maubeuge and

---

[1] Thus Beaulieu in 1796 had to retire from the Boccheta Pass when his subordinate was beaten at Montenotte.

[2] See Arnold, *Lectures on History*, Chap. IV.

Charleroi completed the defence of the Sambre as far as Namur.

Some frontiers are neither mountains nor rivers, but mere arbitrary delimitations, sometimes across plains, in other cases, as in Canada, across every variety of geographical accidents. Between New Brunswick and Maine the treaty frontier, while obtruding this latter State into the eastern part of the Dominion, has not facilitated the strategy of the United States, because the district between the frontier and the St Lawrence could easily be turned into a good imitation of the lines of Torres Vedras. It is true that the Canadian Pacific Railway from Halifax to Montreal passes through Maine, and that this is a serious strategic flaw, but it is not fatal, inasmuch as the Inter-Colonial Railway from Moncton to Quebec runs altogether through Canadian territories on the right bank of the St Lawrence. From Montreal westward to the head of Lake Superior the boundary is water, on which, if the old treaty be not torn up from sentimental reasons, the British have a right to keep a superior naval force. From Lake Superior to Vancouver Island the boundary is a parallel of latitude, in great part over a vast plain rich in cereals and cattle, and partly over the auriferous mountains and beautiful valleys of British Columbia, whose rivers teem with fish. There is no physical or artificial obstacle on either side which could stop a raid against the property and railway-lines of either belligerent for a thousand miles. The boundary between the United States and Mexico is also quite artificial, and though the country contiguous thereto is full of awful natural phenomena —wildernesses, the "llano estacado," wild gorges, the

tremendous cañons of the Colorado, the dismal banks
of the Rio Grande, and the Valley of Death—yet none
of these is of strategic importance. The troops of the
United States in 1846 and 1847 had little difficulty in
marching to the capital of Mexico and winning Texas,
but this campaign was admirably organized and ably
conducted.

The mere geometrical formation of a frontier may
be of the utmost strategic significance. It may be
re-entering as against an enemy whose base is far away;
or it may be re-entering as against a belligerent whose
base is within the angle.

In 1800 Moreau's base was from the Lake of Constance by Basle to Strasburg. The Austrians under
Kray, acting concurrently with their fellow-countrymen
under Melas in Italy, proposed to invade France between
Strasburg and Basle, their line of communication being
from the Black Forest to Vienna. Moreau, by advancing
his right from Schaffhausen, compelled Kray to form
front to flank, beat him at Engen and Mooskirchen, and
compelled him to retreat to Ulm. When the defensive
base is with the re-entering angle or double re-entrant
occupied by the offensive there is no strategic base.
Thus in 1862, the Federals had command of the sea in
the Chesapeake Reach at the mouths of the James and
Potomac, they were also at Fredericksburg, Manassas
Junction, Harper's Ferry, and Franklin, yet they did
not in any case succeed in driving the divisions of the
Confederates from Richmond; in so far as they were
successful at all they drove the Confederates towards
Richmond[1]. A frontier, part of which has an oblique

[1] See map. p. 27.

direction with regard to the line connecting the defensive with its base, offers advantages to the invader. Such was the line from Sierck on the Moselle to the mouth of the Lauter in 1870, by selecting which the Germans threatened McMahon's line of retreat from Woerth to Nancy, and Bazaine's and Frossard's line back to Metz.

The frontiers of Prussia, Austria, and Russia are most interesting from a strategic point of view. Leaving out any considerations of mountains and rivers it will be obvious that Austria-Hungary projects eastward between Roumania and Russian Poland, and therefore threatens the right flank of every Russian movement from the Pruth towards Buda Pesth through Roumania or Servia or over the Danube into Bulgaria. Russia felt this severely in 1854. Again, east Prussia threatens the right flank of any Russian movement through Poland towards Berlin. Further, let us suppose that Austria-Hungary and Germany were allied against the Russian, the latter, when moving against either foe, would be flanked and in great danger from both; accordingly every possible strategic precaution is taken by all three Powers and there are great works and places of arms at Königsberg, Dantzic, Thorn, and Posen in Prussia; Brest-Litovski, Warsaw, Novogeorgovitch and Ivangorod in Poland; and Cracow and Przemysl in Galicia.

Switzerland projects from the French frontier so as to make a salient as between South Germany, Austria, and Italy. In the event of a war between France and the Triple Alliance the possession of Switzerland by either side would have decisive results. If the Swiss joined the Triple Alliance and the French invaded northern Italy by the western Riviera or Mt Cenis, or by a disembarkation at Genoa, and moved on Milan or Venice or Verona, their left would be threatened throughout. If the Swiss joined the French, suppose an invasion of France through the Gap of Burgundy were proposed, or even by way of Brisach or Strasburg, the left of the Germans would be threatened, and a combination of the Austrians and the Germans by way of the road south of the Danube would be difficult, while any movement of the former through the Black Forest to the Rhine would be perilous. Switzerland therefore is interesting not only for the illustration of mountain fighting in its own territory, but because of the relation between its river valleys and passes and European strategy at large. But the Swiss are taking precautions to prevent their country being made an avenue for contending armies ; nor are they likely to allow any philosophy of the rights of man to beguile them again into giving ambitious republicans a military foothold within their borders, as was the case with the French in 1799.

The frontier of Italy has serious defects. On the west not only has she the territory up to the watershed, but the heads of the valleys of the French slope are in her hands. On the north it is clear that many of the issues are commanded by other States. A zigzag frontier, however mountainous, is defective as exposing flanks, but a zigzag frontier when the upper reaches of the river avenues are in command of an opponent is very unsatisfactory, and in mountainous countries rivers and valleys and roads are almost synonymous. On the north the upper valleys of most of the tributaries of the Po and of the Adige are not Italian. The Swiss canton of Ticino and the Austrian Tyrol obtrude into Lombardy. In the north-east also Austria still commands all the mountain issues.

All the roads leading from Italy into Switzerland converge in the longitudinal *couloir* from the Rhône to the Rhine, from which there are only three exits—that of the Rhône closed at the defile of St Maurice, that of the Rhine closed at the defile of Sargans by the old fort of Luziensteig, where new works are to be constructed, and that of the Reuss commanded by the new works of St Gothard. St Maurice is a strategic point of the very first consequence ; it is one of the best positions for a fortress in Europe, and properly fortified and occupied by a brigade is practically impregnable, and absolutely closes the valley of the Rhône near Sion. So important is this defile, the key of the Valais, that the occupation of the Simplon and of the Great St Bernard would only be a temporary success for the Italians, unless they seized it by a *coup de main*. In consequence of a series of bends or zigzags similar to those on the northern Italian frontier, the river Main cannot be regarded as a line of defence, and cannot protect Bavaria from attacks from the north. The South Germans in 1866 made a great mistake in using the middle and lower Main. Had they moved towards Saxony by the upper Main, as did Napoleon in 1806, and concentrated in Thuringia, they would have seriously hampered the Prussians in their operations against Austria.

The present Franco-German frontier has towards France the form of a great *tenaille,* the exact meaning of which is the re-entering angle at the point whence issue the Meurthe and the Saar, behind which are the vast works of Strasburg. The southern Vosges strengthen the southern side, while Metz and Diedenhofen cover the extreme right on the northern flank. The extreme left is not very strong by nature, and allowing the French to retain Belfort, however judicious from a political point of view, was a glaring strategic error. But the proximity of Strasburg and Brisach, and the Rhine itself, together with the fortifications of Mulhausen and Altkirch, counteract the weakness of this flank to a considerable degree. On the whole a French movement against this *tenaille* would be very hazardous. In the event of a war between France and Germany, without allies on either side, the front of operation must be the line from Porrentruy to the boundary of Luxembourg. The neutral territories of Belgium, Luxembourg, and Switzerland protect the extreme flanks of the belligerents and would bar any wide strategic turning movement. The French attack must at first be a front attack. The best line of advance would be Paris—Mannheim, but it would be almost suicidal to push the advance far, having regard to the fortresses on each flank and the difficulties of the hilly country between the Moselle, the Saar, and the Rhine, with numerous retarding positions. Moreover, these once surmounted, in front would be the Rhine itself, the banks of which, assailed in the past by a thousand battles, are to-day the lines of two strategic railways and are commanded by powerful fortresses which must be taken or masked by many corps. On the assumption of a French initiative the new frontier resulting from the loss of Alsace and Lorraine, quite apart from the relative strength or efficiency of the contending armies, has turned the balance to the distinct disadvantage of France.

The timidity or the prudence of statesmen entrusted with the care of the frontiers of nations or of distant provinces has often induced them to construct boundaries or inner defences in the shape of walls, series of bastions, curtains, and lines. Hadrian, Trajan, and Antoninus were celebrated builders of walls in many parts of Europe from the shores of the Black Sea to the banks of the Clyde, but the ability of the engineers and expenses of the exchequer were in vain. If long reaches of sea, broad rivers, and intricate mountain districts cannot protect a state, neither can a lengthy wall. A few good fortresses in determined hands on rivers near the frontier would be much better. The lines of Torres Vedras covering Lisbon will be described in another place ; they effected their purpose completely, and their construction was justified by the topography of the area of operations, and by the fact that each flank was rendered invulnerable by sea power. But the long lines of works constructed in Belgium in 1705, from near Namur to Antwerp, were carried by Marlborough. This ought to have been a lesson to other French generals, but Marshal Villars, as has already been stated, constructed even more formidable works for the defence of France in 1710, which he said would cause Marlborough to reach his *ne plus ultra*. The folly of such works, however, was again proved by the ease with which Marlborough got through them. This kind of artificial frontier is useless, and has always been forced when skilfully attacked.

The military incompetence of the Chinese has been

recently illustrated in a striking fashion by the passive
nature of their strategy, which relies on anything rather
than powerful, mobile, and well-equipped armies.   They
prefer any measures, however laborious and costly, rather
than a good fight in the open field.   Their Great Wall
is a massive and marvellous record of incompetence.
It is carried over the highest ranges and across rocky
barriers with defiles in which small bands of brave men
could easily stop multitudes without any artificial works.
With its windings, and double and triple lines at certain
points, it is 2000 miles long—the distance from the
Rhine to the Ural Mountains, and from the Straits of
Gibraltar to the Crimea.   Yet it could never keep out
invaders, even if maintained as carefully as a Parisian fort.
It is not only the most prodigious example of strategic
ineptitude but it is also a sad specimen of official negligence and corruption.   As Captain Younghusband tells
us, the inner branch north of Peking, where it is "under
the eye of the Emperor, is a magnificent structure built
of immense blocks of granite.   It is some 40 or 50 feet
in height, and wide enough at the top to drive two
carriages abreast, winding up and down the steep hill-
side over the summit and across the valleys far away
into the distance.   But where I passed it next, scarcely
one hundred miles from Peking, it had dwindled down
to a miserable mud wall, not 20 feet in height, of no
thickness, and with gaps in it from a quarter to half a
mile in width.   At the gateway were massive doors
and a lofty gate, guns pointing down the road, and a
detachment of soldiers to collect the customs dues, while
twenty yards to the right was a gap in the walls wide
enough for a brigade in line to pass through."   Yet the

Chinese Empire was most richly endowed with all the conditions of national greatness. Its decay is only the most appalling illustration of the axiom that no nation can be great that does·not make the art of war its principal study. Excluding the Pamir, a kind of neutral land "where three empires meet," the central plateau of Asia where rise its mightiest rivers is under Chinese rule, and its boundaries are clearly defined on the north and north-west by Asiatic Russia, on the south and south-west by British India, on the south-east by Indo-China, and on the east by the Pacific. Inside these boundaries there are 4,500,000 square miles and 350,000,000 people. The agricultural and mineral resources of all kinds are practically inexhaustible, and yet the whole nation is a prey either to a small neighbouring state like Japan, or to distant European countries like Germany and England. And why? Because the military profession was despised by the adherents of a humanitarian philosophy, and for no other reason. The Chinese make very good soldiers when properly drilled and led by warriors like General Gordon.

In fine, with regard to boundaries, we find that certain frontiers give military chances to brave and enterprising armies, and that no frontier, however massive, or however strengthened by nature and art, can prevent a luxurious, inert, or corrupt race from ruin.

Readers who wish to study at large the use of practical geography in the determination of frontiers, and the vast losses to our wealth and prestige which have resulted from official ignorance of it, should read Colonel Sir T. Holdich's paper published in the *Geographical Journal* for May, 1899. Our rulers have been led astray in every continent by the neglect of the study of modern geography in our schools and colleges. Mercantile and railway routes have been spoiled by superficial or careless surveys, and intricate international complications have arisen which even a limited knowledge of geography would have prevented. Sir Thomas Holdich says :—"This period in our history has been well defined as the boundary-making era. Whether we turn to Europe, Asia, Africa, or America, such an endless vista of political geography arises before us, such a vast area of new land and sea to be explored and developed, such a vision of great burdens for the white man to take up in far-off regions, dim and indefinite as yet; that it can surely be only by the grace of Providence that we shall finally emerge from the struggle to rearrange the world's partitioning without some deadly contest with others whose interests in these new arrangements are hardly less than our own. And I may perhaps be permitted to say, that just as the Providence of battles usually favours the biggest battalions, so it is likely that the widest geographical knowledge will prove the best safeguard against misunderstanding, and will at once dispose of such false estimates of the value of portions of the world's surface here and there as have occasionally brought England perilously close to the dividing-line between peace and war. By geographical knowledge I do not mean simply that knowledge of the earth's surface which we gain by surveying it. I mean also a knowledge of those ordinary laws of nature which decide the configuration of mountains and the flow of rivers, where certain influences must inevitably lead to certain conditions. I mean, also, such knowledge of the technical application of geographical terms as will prevent mis-understanding about the meaning of words and phrases.

"Of all sources of international irritation, boundaries seem to be the most prolific ; and of all countries in the world, England has probably suffered the most from them. To refer to modern history only, it was the Sistan arbitration which first turned Sher Ali's heart against us and originated the Afghan war of 1879–80 ; it was a boundary which brought England and Russia face to face in Turkestan in 1884, and so nearly forced us into war; it was a boundary (nothing less) that started Umra Khan on his quest for Chitral ; it was a boundary which set all the north-west frontier in a blaze lately. And yet all this boundary-making has been in the interests of peace alone. The want of these boundaries would more surely have led to wider-spread, more disastrous war than the making of them, and it seems of all things most extraordinary that efforts honestly made in the interests of peace and good government should not be possible, without bringing great countries to the verge of blows."

In arranging frontiers a wrong definition may lead to irritating disputes, if not to actual war. To quote Sir T. Holdich again :—" Supposing it were a matter of determining a boundary between India and Tibet. Standing back some 100 miles from the plains of India, in the centre of the Himalayan mountains, is a magnificent central watershed, or water-parting, which stretches from Kashmir to northern Assam. The greatest snow-peaks and glaciers of the world are piled on to the summits of this vast crystalline axis of the Himalaya. Could anything be better than this magnificent array of unapproachable

snow and ice to serve as the unmistakable barrier between two vast Asiatic countries? Nothing could be
better, provided we do not define it as the watershed
between India and Tibet. From its southern flanks the
first beginnings of many mighty rivers flow southward
to the plains of India; from its northern buttresses and
spurs many a torrent pours northward, and turns equally
to the plains of India. The Indus and the Brahmaputra
drain the northern slopes of the central Himalaya,
enclosing the great mountain system between them,
whilst the largest affluent of the Ganges cuts it right in
two.

"What is true of the Himalaya is true of nearly all
the great mountain systems in the world, *i.e.* the watershed of the system is beyond, and apart from, the highest
mountain chain. This is true also, in a smaller sense, of
smaller ranges, so as to make it essentially necessary
to distinguish between a central chain of peaks and the
water-divide of the system as a whole."

CHAPTER IX.

FORTIFICATION AS RELATED TO MILITARY GEOGRAPHY.

A WRITER on military history is fortunate indeed if he can escape all the controversies which perplex both the theorists and the practical engineers who discuss the various systems of fortification. Though tactical principles and human nature are practically the same in all ages, the only result of their study when applied either to the fashion or the position of fortresses seems to be a perfect riot of disagreement and confusion. Many would turn a land fortress into an imitation of a ship, and apply cupolas and all kinds of machinery such as are common in monster ironclads to subterranean forts.

General Brialmont is an ardent advocate of *batteries cuirassées*, but the Germans look at his schemes askance; and Colonel G. Sydenham Clarke ridicules all elaborate systems, and entertains a profound contempt for the whole art of permanent fortifications. Our own permanent land defences he holds in small esteem. "Our modern works both at home and abroad might have been designed by clever cadets, quick to recognise the niceties of technical artifice, but unable to grasp the broader aspects of the science of war." An indifferent and badly handled force cannot be made safe by money and art. "The best fortification, judged by results, has been that improvised by stress of circumstances, unspoiled by the debasing influence of the text-book, and not demoralised by the technical possibilities opened out by large expenditure."

The question of the utility or otherwise of the vast expenditure on French fortresses since 1870 is most interesting and instructive. Other nations would do well to hesitate before embarking on a similar career of sinking capital in bricks and mortar. Pierron believes that most of them are utterly useless, if not dangerous. One of the great inconveniences of every fortress is that it shuts up behind walls a certain number of men, say from one to twenty thousand, who might be much better employed with the armies in the field. The French fortresses would altogether shut up no fewer than 600,000 men, or more than the total of both French and German armies at Gravelotte. So, in 1814, a large portion of Napoleon's best troops were wasted in German fortresses, when they would have been invaluable in Champagne. Worse still, many authorities contend that the sites of the new works have been very badly chosen, and some writers explain this by the fact that the new system of defence has been entrusted to Engineer officers. Of course the task of constructing works rightly appertains to these officers, but they are in no sense superior to others in ability to select suitable sites. Indeed they are less capable, as their education causes them to look at things from the point of view of the art of fortification only, rather than from that of the art of strategy.

Another vital question on which authorities differ is the fortification of capitals and leading towns of a State. Pierron contends that these should never be fortified, that their investment is disastrous from every point of view, military, political, social, and commercial. He would immediately destroy the forts of Paris and Lyons, and in their stead erect defences in places of trifling wealth and small political importance, which if on the flank of the enemy's line of advance to a capital or great commercial depôt, would protect it as effectively as a girdle of forts. On the other hand, most writers, including Marmont and Hamley, insist on the fortification of the centres of national wealth and culture, and schemes for the fortification of London itself are frequent in military societies, but a complete investment of London would be a stupendous enterprise. Since the erection of the new works at Paris, the number of army corps required to isolate it as effectually as was the case in 1870, would be enormous.

It is an incontestable fact that no kind of fortress, wheresoever placed, however strongly manned, however expensively constructed, and however numerous its garrison, has ever given permanent security to a State— has seldom indeed given it even temporary protection. Moreover, a fortress once invested is certain to fall, unless a relieving field-army can beat the besiegers away. We read in the history of one generation of the "virgin" fortress of Ingoldstadt or of Metz, but when we open the records of another generation, we find that its pride has bitten the dust.

Belgium bristled with fortresses in the time of Marlborough, but sometimes a dozen surrendered after a decisive battle like Ramillies, and the illustrious Englishman was able to pass the French frontier, to take Lille after one of the most celebrated sieges in history, and threaten Paris itself. After the battles of Jena and Auerstädt, Napoleon crossed the Elbe and entered Berlin irrespective of Magdeburg and Dresden. The French frontiers north and east were in 1814 well covered with works illustrating the genius of Vauban; nevertheless, the only hope of France was Napoleon's army on the Seine and Marne. The allies from the Rhine and Belgium did not waste their time in sieges; when once Napoleon's energy in the field was exhausted, Paris was entered and peace dictated. So, after Waterloo, the northern fortresses of France scarcely delayed the march of Wellington and Blucher on Paris. The Germans won the battle of Weissenburg on the frontier August 4th, 1870; they were around Paris September 19th, a distance of 250 miles, in spite of the fact that the French were in occupation of the vast places of Strasburg and Metz, and many an intervening fort. These are cases in which a country was traversed by an enemy as he pleased, although many fortresses were still intact and strongly garrisoned. It is true that Plevna hampered the Russians, and the works of Richmond and Petersburg delayed the Federals for considerable periods, producing a decisive effect on the war, but both were certain to fall in due time if no field-army came to their aid. They fell at last; their defenders, in each case numerous, became prisoners, and the tide of conquest rolled on.

There are, however, some cases in which fortresses, well constructed and with moderate garrisons, give security to this extent—that till they are taken the enemy's further advance is impossible. This is when the avenues into a country are few, perhaps one or two main roads only. If a strong place with a resolute commander and a well-found garrison covers each of the roads, the invader cannot advance till he takes the fortress. Such were Ciudad Rodrigo on the road from Almeida to Salamanca, Badajoz on the Guadiana, and San Sebastian and Pampeluna on the Pyrenean frontier. Wellington was stopped in 1811 by the two former, but he regularly besieged, breached, and stormed them, in spite of the most splendid resistance, in 1812. San Sebastian and Pampeluna blocked his way from the battle-field of Vitoria into France in 1813. He therefore breached and stormed the former, and blockaded the latter, before daring to wend his way towards the Adour. For lack of assistance from outside, therefore, these most valuable places, all defended *à outrance*, were captured. Marmont and Soult had armies quite strong enough to beat off the besiegers in 1812, but they were surprised and had no time to act effectively. Soult made bold efforts to save Pampeluna and San Sebastian, but was routed in the passes of the Pyrenees.

In some cases a very small fort in a well-chosen position may puzzle a general of genius. Bard blocked Napoleon's way as he descended from the Great St Bernard towards the plain of Lombardy, but he easily evaded its guns by a devious path and simple stratagems. There are certain portions of territory in many states which, if resistance to the bitter end be resolved upon, might be defended for an indefinite period against almost any force. Such is the peninsula of which Constanti-

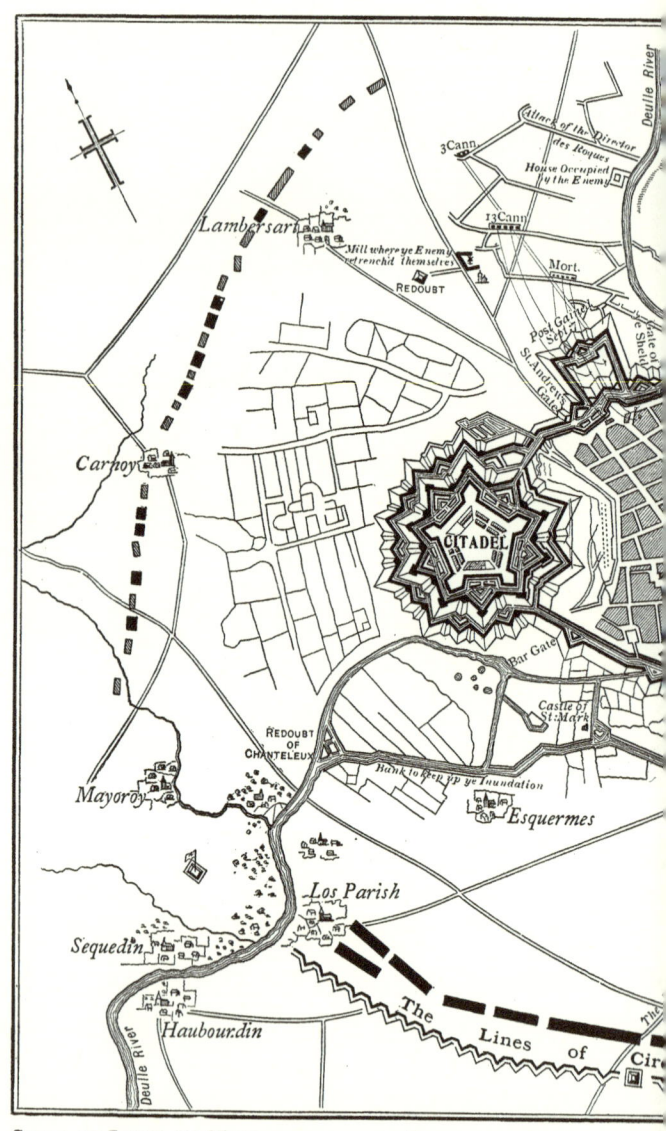

SIEGE OF LILLE BY MARLBOROUGH AND EUGENE IN 1708.

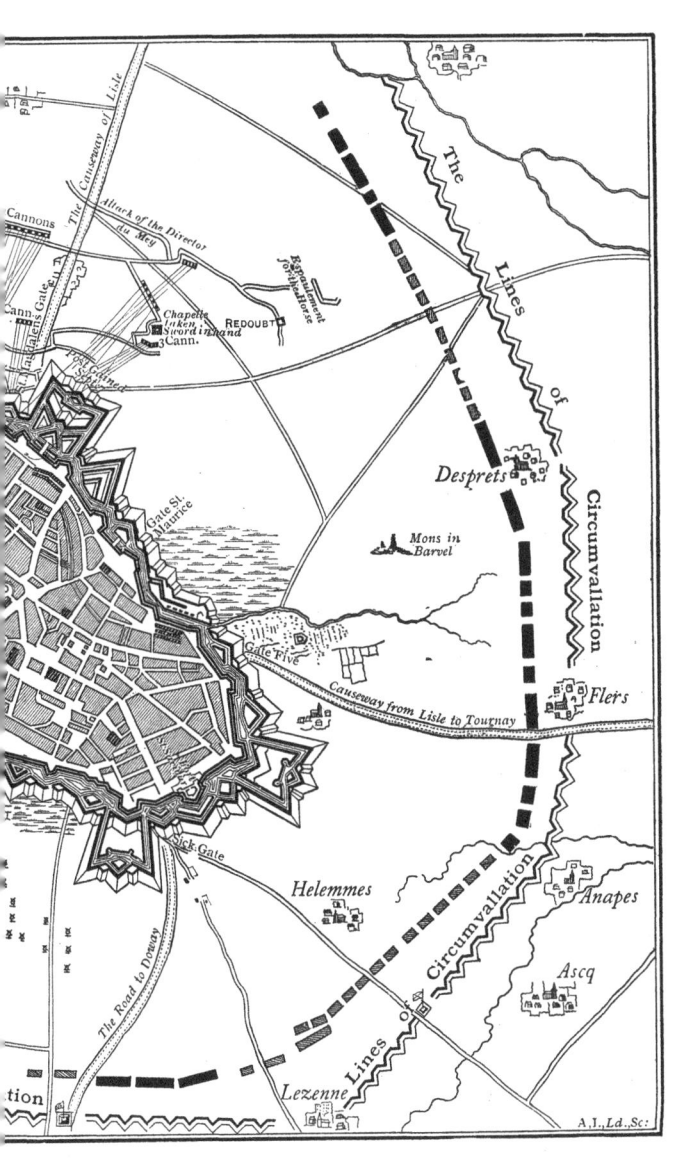

The Causeway of Lisle

Cannons

Attack of the Director du Meg

Fagnia's Gate

Redoubt taken Sword in hand & 3 Cann.

Chapelle taken Sword in hand

Redoubt

Emplacement for the Gate House

Cann.

The Causeway of Lisle

Gate St. Maurice

Desprets

Mons in Barvel

Gate Five

Causeway from Lisle to Tournay

Flers

Dick Gate

Helemmes

Circumvallation

Anapes

The Road to Doway

Ascq

Lezenne

Lines of Circumvallation

...tion

A.I.,Ld.,Sc:

The Lines of Circumvallation

nople is the apex. Lines were constructed in 1877
north of Constantinople from sea to sea which brought
the Russians to a standstill, especially as the fleets of
Admirals Hornby and Hobart Pasha commanded the
sea on every side. The Russians were soon hard pressed
by want of food and by disease. To go on was impossible ; to march homeward over ruined Roumelia and
Bulgaria across the carcass-strewn passes of the Balkans
to their bridge on the Danube, and thence by Roumania
would have meant annihilation. No wonder that they
welcomed an armistice and food by sea ; otherwise their
case would have been far worse than was that of the
astute Massena when, after long observation of the
fortified Lisbon peninsula, he sullenly began his retreat
from the first of the lines of Torres Vedras through
devastated Portugal back to the Agueda and Salamanca
in 1811. The position of Buyuk-Tchekmedje is as remarkable in natural military strength as is Constantinople in geographical situation. The peninsula (between
the Black Sea and the Sea of Marmora) is here but
twenty miles wide, and twelve miles of this space are
occupied by broad lakes extending up inland from either
shore. Of the remaining eight miles at least half is
filled with impassable or difficult swamps, and the
rest with almost impenetrable thickets. Behind this
line of lakes, swamps, marshes, and thickets runs a
continuous ridge from sea to sea, from 400 to 700 feet
in height, and on this ridge the Turks had in process of
construction not less than thirty large redoubts, besides
outlying trenches and rifle-pits, the greater part of them
concentrated in the centre of the line, and disposed
irregularly, according to the nature of the ground, in

three lines. These redoubts were only half finished, but still they afforded complete protection for infantry, they would have mounted 150 siege-guns and as many more field-guns, and their proper garrison would have been 60,000 to 75,000 men.

With such a garrison, since the flanks of the line rested on the sea, and could not be turned or invested, these lines might fairly be called impregnable. The force actually in them consisted of about 30,000 men, made up from the wrecks of Suliman's army, which had been brought by sea from Enos; of Achmed Eyoub's division, which had retreated from Adrianople; and of some reserves which had been at Constantinople during the war, the whole being under the command of Ghazi Moukta Pasha, who had lately returned from Asia, where he had lost his whole army. Yet such was the natural strength of this position, taking into account the shortness of the line, which allowed the men to be within easy supporting distance of each other, that 30,000 men here constituted a more formidable adversary than 60,000 men in the line of works held by Osman at Plevna. But the armistice gave these away with a stroke of the pen to the Russians.

Wellington's campaigns of 1810 and 1811 were a model of defensive warfare, the result of most careful study of the strategic geography of the theatre of operations, and have received the warmest commendations of French writers; even as Massena and Ney, though repulsed, got a certificate from their great opponent to the effect that he was never free from care and danger when they were near. The fortifications on the frontier, Ciudad Rodrigo and Almeida, could not save Portugal,

nor could the British army fight in the open field a
series of combats against the forces of the "spoiled child
of victory," superior in everything except courage, and
not deficient in that. So Wellington cautiously retired
from the frontier, leaving Craufurd with a rear guard to
delay the invader. When he came near the Mondego he
halted at Busaco, and received and repulsed the fierce
attacks of his pursuer. Having won a battle, he continued
his retreat. Massena followed. But in the previous
winter, with infinite pains and complete secrecy, the celebrated lines of Torres Vedras had been constructed,
and accordingly, when Massena fancied that he was
driving the British into their ships, he found that his own
march was stopped by strong redoubts and formidable
abattis. As at this time the British had uncontested
command of the sea, there being no other European
"fleet in being," both flanks were absolutely secure, and
the defenders could be abundantly supplied by our fleet ;
while they had all the resources, mechanical and otherwise, of a large and flourishing capital city at their
disposal. Moreover, the Baragueda spur of the Monte
Junto came down in a perpendicular direction close to
the works of the first line, and was a parallel obstacle
with no transverse passages, which narrowed the French
sphere of attack on the first line to fourteen miles on
the Tagus side ; nor did they dare to divide their forces
by approaching the lines on both sides of the spur.
Manifestly in this case, from a broad point of view,
strategic geography was with Wellington.

As the lines of Torres Vedras are the finest and
most successful example of field work of this kind in
history, and indeed have had no rival since the days

of Trajan, since they utterly ruined the enterprises of
several of the best warriors of our century, it may be
well, ignoring all technical details, to set forth their
topographical conditions as described by Jones and
Napier. A glance at the map will show that both
flanks were covered, on the one hand by the sea and on
the other by the Tagus, and could be protected by our
navy.

They consisted of three distinct lines of defence.

The first, about thirty miles north of Lisbon, extending from Alhandra on the Tagus to the mouth of
the Zizandro on the sea-coast, and following the inflec-
tions of the hills, was twenty-nine miles long.

The second, traced at a distance varying from six
to ten miles in rear of the first, stretched from Quintella
on the Tagus to the mouth of the S. Lourenço, being
twenty-four miles in length.

The third, intended to cover a forced embarkation,
extended from Paço d'Arcos on the Tagus to the tower
of Junquera on the coast. Here an outward line,
constructed on an opening of three thousand yards,
enclosed an entrenched camp, the latter being designed
to cover an embarkation with fewer troops if such an
operation should be delayed by bad weather.

Of these stupendous lines, the second, whether for
strength or importance, was the principal, the others
were appendages, the third being a mere place of refuge.

Five roads practicable for guns pierced the first line
of defence; two at Torres Vedras, two at Sobral, one
at Alhandra; but as two of these united again at the
Cabeça, there were only four points of passage through
the second line.

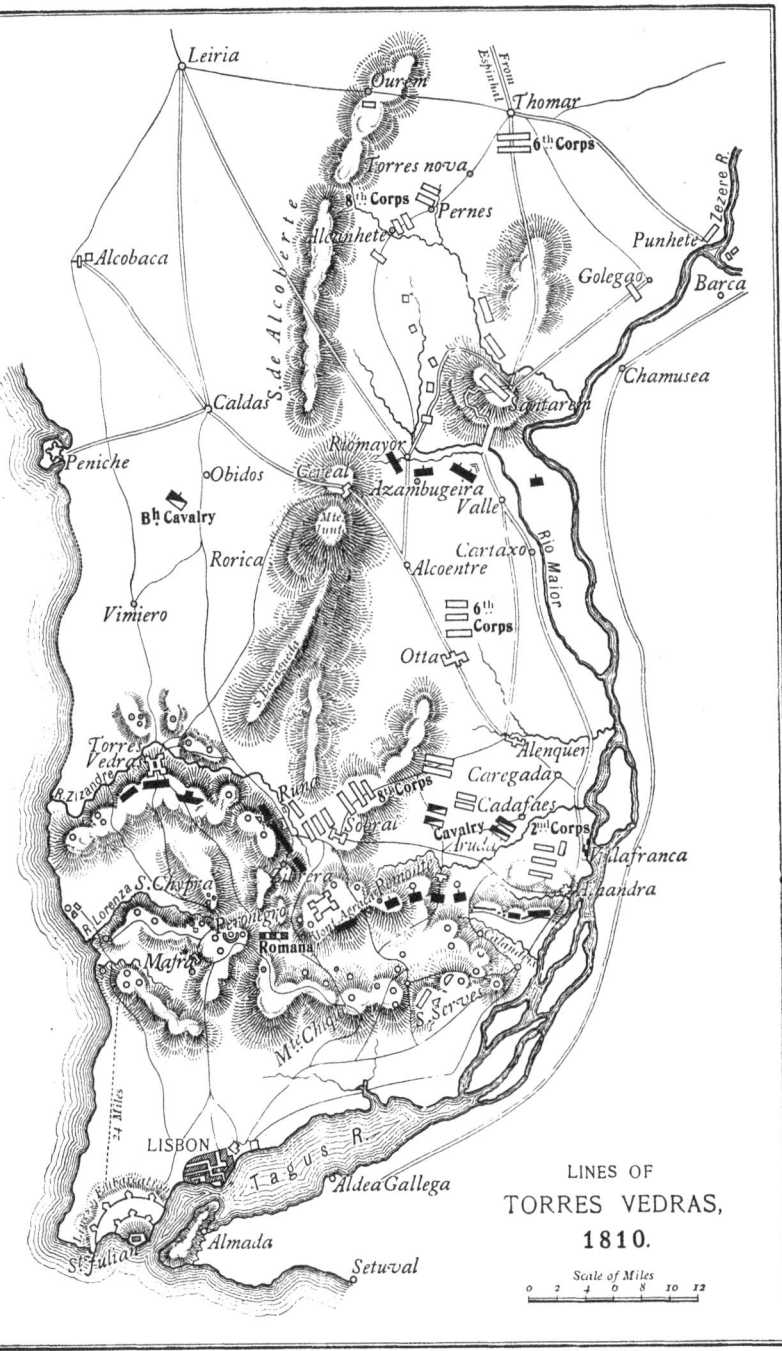

Leiria

Ourem

From Espinhal

Thomar

6th Corps

Torres nova

8th Corps

Pernes

Alcanhete

Alcobaca

Punhete

Golegao

Barca

Santarem

Caldas

Rhomayor

Chamusea

Peniche

Obidos

Cereal

Azambugeira

Valle

B^h Cavalry

Rorica

Cartaxo

Alcoentre

Vimiero

6th Corps

Otta

Alenquer

Torres Vedras

Caregada

R.Zizandre

Sobral

Cadafaes

9th Corps

Cavalry

2nd Corps

Villafranca

S.Chypra

Arruda

R.Lorenza

Pero negro

Aveiras

Alhandra

Romana

Mafra

M^{te}Chiqu

S.^a Serves

LISBON

LINES OF
TORRES VEDRAS,
1810.

Tagus R.

Aldea Gallega

St Julian

Almada

Setuval

Scale of Miles
0   2   4   6   8   10   12

Hence the aim of all the works was to bar these roads and strengthen the favourable fighting-positions between them, without impeding the movements of the army. The loss of the first line, therefore, would not have been injurious, save in reputation, because the retreat was secure upon the second and stronger line; moreover the guns of the first line were all of inferior calibre, mounted on common truck carriages, immoveable, and useless to the enemy. To occupy fifty miles of fortification, to man one hundred and fifty forts, and to work six hundred guns required many men, but numbers were not wanting. A great fleet in the Tagus, a superb body of marines sent out from England, the civic guards of Lisbon, the Portuguese heavy artillery corps, the militia and ordnance of Estremadura, furnished a powerful reserve to the regular army. The native gunners and the militia supplied all the garrisons of the forts on the second, and most of those on the first line; the British marines occupied the third line; the navy manned the gunboats on the river, and aided in various ways the operations in the field.

Near Aruda a loose stone wall, sixteen feet thick and forty feet high, was raised; across the great valley of Aruda a double line of abattis was drawn; not, as usual, of the limbs of trees, but of full-grown oaks and chestnuts, dug up with all their roots and branches, dragged by main force for several hundred yards, and then reset and crossed so that no human strength could break through. Breast-works, at convenient distances to defend this line of trees, were also cast up; and along the summits of the mountain, for a space of nearly three miles, including the salient points, other stone walls six feet high by four in thickness, with banquettes, were constructed.

The success of the British at Torres Vedras misled other generals, and the Russians tried a similar structure at Drissa in 1812, which was a complete failure. But Soult's defensive works in the south-west of France in 1813 were like the British lines inverted, the apex being Bayonne and the base a line from St Jean Pied de Port to the mouth of the Bidassoa. Inside this base the Nivelle was strongly protected, so as to make its passage a serious undertaking, and the Nive was a parallel obstacle to the British advance, across which Soult manœuvred very ably against each British wing alternately. Bayonne formed a very useful *entrepôt*, the investment of which was hampered by the Adour. The value of Soult's precautions is proved by the fact that whereas the battles of the Pyrenees were over by August 2nd, 1813, Bayonne was not invested till January 14th, 1814. It had not fallen when Napoleon abdicated; indeed an unusually successful sortie took place April 14th, some days after that event. There can be no doubt that Soult's plans for France were inspired by recollections of the British defence of Portugal.

References to the "Lines" are frequent in all works on strategic geography. American strategists set great store on them when considering plans of defence and military topography, especially when land is supported by sea power. The defence of Canada has recently occupied a considerable share of attention on both sides of the great lakes. Let us suppose a United States force rapidly concentrating in Maine with the intention of invading the eastern provinces, beginning with New

M.   13

Brunswick.   It is the opinion of Captain Wagner, of the United States Army—a good authority—that it would soon find itself face to face with a transatlantic Torres Vedras.   The main force of the British would doubtless retreat behind the Petitcodiac River[1], where it would find a position of remarkable strength.   With a front of less than 15 miles, the army could rest its right flank on Northumberland Sound and its left on the Petitcodiac, a great part of its front covered by a small river, and a railroad running along the rear of the position. The flanks could not be turned, the navy could deliver a flanking fire along the lines, reinforcements could be speedily sent from one part of the line to another, and supply from the sea and by rail from Nova Scotia would be sure and easy.   Unless this territory presents disadvantages not hinted at by any ordinary map, it would be a position not one whit inferior to the Lines of Torres Vedras ; for while there would be no Monte Junto to divide the assailant's front, the position is scarcely more than half as long as Wellington's famous lines, the navy (from the nature of the position and the increased power of its ordnance) could lend a greater degree of assistance than it was able to afford the Iron Duke, and the railroad would give advantages not dreamed of 80 years ago in Portugal.   St John would be connected with the lines by about 120 miles of waterway, by means of which its garrison could be withdrawn to the lines of Petitcodiac, should the New Brunswick metropolis prove untenable.   Should the freezing of the river deprive the line of the support of the navy, and thus expose the flank, a position of almost equal strength could be taken

[1] See sketch-map.

up 30 miles to the rear, where the army would form on
about the same front as before, its left resting on Cumber-
land Basin, its right on Bay Verte.   On the left is a high
hill, occupied in colonial times by the French fort
Beauséjour, while a great portion of the front is covered
by marshes.   In fact, the absence of the railway in rear

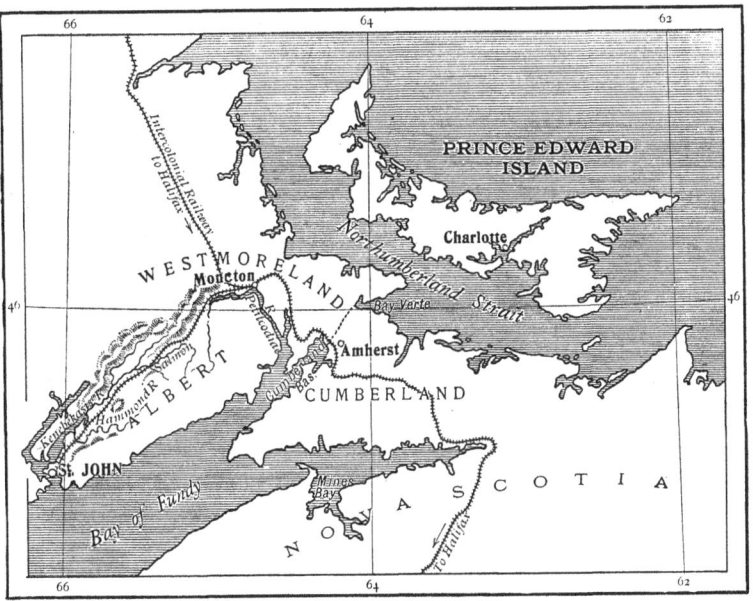

ENVIRONS OF THE PETITCODIAC RIVER.

of, and parallel to, the lines is the only point in which
this position is inferior to the one on the Petitcodiac;
while the marshes covering its front would make it much
stronger as a purely defensive position.

In the late wars Cuba offered admirable opportunities for guerrilla warfare—better than Galicia, the Basque
provinces, and Catalonia afforded the mother country

in Napoleonic days. Gomez and Maceo recalled to memory the exploits of Mina and Porlier, and Lacy and O'Donnell, and a singular feature of the two great Cuban wars against Spain, 1868—1878, and 1895—1898, was that the regular troops of the latter adopted the tactics of their adversaries, only that they supplemented a petty *partidas* warfare by antiquated and utterly futile systems of lines and forts. The Cubans never committed themselves to an action on a large scale, much less to a decisive conflict; and the Spaniards, though they put 200,000 regulars and 60,000 volunteers in the field as against some 40,000 to 60,000 insurgents, never realised the necessity of opening up roads into the enemy's country, and following up the enemy to his lairs, though they had plenty of examples in the fashion in which Suchet, Soult, and Reille acted in their own country.

The centre of Cuba is traversed by a mountain range, broken at intervals and of an average height of 2,200 feet. The hills are wooded, and the valleys full of forests and jungles in which long grass abounds. After a skirmish the insurgents retired into these recesses, and from ambuscades therein surprised hostile forces on their marches. The Spanish system at the opening of a campaign was to convey a force by land or sea to a point near which the insurgents were supposed to lurk, then drawing up the force in narrow columns, to cut a pathway into the forest. If the insurgents were met a few shots were fired on each side; then the troops retired or the rebels made off to some new fastness.

Meanwhile, death from disease was raging among the Regulars. In the 1868—1878 war 75,000 died. On the other hand, it was impossible to starve the insurgents in what Christopher Columbus described as "the most beautiful land that the sun ever shone upon, or the eye of man ever beheld." Yams, sweet-potatoes, and bananas abounded, and a crop yields food in plenty two months after being planted. If the Spanish system of entrenchments or *trochas* right across the island was based on an imitation of the lines of Torres Vedras, they entirely misconceived the ideas of the British engineers in Portugal.

The original *trocha* or trench crosses the island between the provinces of Puerto Principe and Santa Clara, where the country has but little elevation above the sea-level; its length was 50 miles. It was flanked by swamps and was strengthened at intervals with anti-quated forts. This trocha was constructed during the insurrection of 1868 to 1878, and Campos rebuilt it in 1895. It was barely finished when both Gomez and Maceo crossed it, and carried the war into Santa Clara. Campos then constructed a second trocha, nearer to his capital, that is, further west, between Las Cruces and Las Lajas, and, this proving equally ineffectual, a third, between Matacezan and La Broa. This latter was only 28 miles long, and was crossed by the railway from Havana to Batabano, on which ironclad freight-cars, pierced with loop-holes for rifle-fire, accompanied each train. Gomez and Maceo, however, tore up part of the track.

When Weyler succeeded Campos he built an additional trocha 25 miles west of Havana, between Mariel and Majana, expecting to shut Maceo up in Pinar del Rio. This consisted firstly of a clearing in the forest,

100 to 800 yards wide; secondly, of a barbed-wire fence, four feet high, behind which were stationed the sentinels; thirdly, forty yards in rear, of a trench, three feet wide and four deep, with a breastwork of logs ; and fourthly, fifty yards further back, of a chain of log-houses, each containing a garrison of 100 men.

Another new trocha, built also by Weyler, extended from Jucaro to Moron, in the province of Puerto Principe. This had no ditch, but, in its stead, a wall six feet high, composed of felled trees, while a military railway extended the whole length of the cleared space. Its forts were of a threefold kind—large ones, half a mile apart, blockhouses midway between these, and small huts for guards of five men each, there being three of these huts to every quarter-mile interval. Entrance to the forts was obtainable only by ladders, which could be lifted and withdrawn from within ; and bombs were distributed at intervals along the lines.

In addition to the forts along the various trochas, and unconnected with them, or, indeed, with any other system of military defence, the Spaniards had an enormous number of small forts all over the island, not less than two thousand, and possibly even more. They proved an additional heavy drain on Spain's resources in men, and we have it on the authority of Consul Lee that the trochas, though costing a large amount of money and absorbing a considerable force of men for their defence, proved quite useless and were in the end practically abandoned.

Spain had absolute possession of the cities, fortified towns, and forts. With the exception of these, that is, in the country at large, the Cubans were masters.

On both sides the workmanship was wretched ; on the part of the Spaniards so much so that the insurgents often ventured quite near to the forts and trochas. The Spaniards rarely left their fortified positions. If they did, they returned in time for dinner, or, in any case, before nightfall. Their system of defences was good enough in its way, but as it was unconnected with any offensive measures, it might as well have been non-existent, for any difference it made in the progress or ultimate results of the war.

The increased range of modern artillery has profoundly modified the situation of fortresses, so much so that the outlying forts must be located further and further from the town or citadel to be defended. Many sites which to old engineers seemed admirably adapted for fortresses are now of no use, and the places erected thereon at a large expense by previous generations are being dismantled. A similar fate has long since overtaken their predecessors, and the old medieval and 16th and 17th century works have been transformed into boulevards. For the same reason it was found in many cases during the Franco-German war that it would be waste of time to go through the tedious process of a siege, and that the various engineering devices could be ignored and a bombardment directed not on the walls, which were supposed to protect the houses, but on the houses themselves. Thus Péronne, Thionville, and Mézières, after a lamentable destruction of property, were obliged to surrender in a few days. The Commandants took pity on the inhabitants. This was against the policy of Faidherbe, who would have held the towns as long as the troops could man the ramparts, irrespective of the sufferings of the people at large, the blame of which he would lay on the assailants.

Admirable sites for fortresses are either hills commanding bridges over which invading armies must pass, or bridges themselves, when no hill is in the neighbourhood. When a river is parallel to the line of advance of an enemy, fortresses protecting the main bridges are most serious obstacles ; they flank his line of march and he must either stop and take them, or detach troops to observe them and thus weaken the army. Thus Napoleon in 1809 was obliged to observe Ulm, Ratisbon, and Passau, as well as Linz and Krems, so that when he came to the island of Lobau and was checked at Aspern and Essling, he was obliged to wait for reinforcements from Italy before making another stroke for victory at Wagram. Had Linz been well fortified, he would have been in a still worse case. Napoleon III. finding that if he advanced from Alessandria on Piacenza he would be exposed to the works of Stradella, changed his line of operations from the south to the north of the Po previous to the battle of Magenta on the Ticino, in 1859.

If several rivers converge, and other circumstances are equally favourable, their points of convergence are admirable sites for a fortress. Metz is at the junction of the Seille and Moselle, and with its outlying forts on the hills was almost impregnable in 1870 ; it fell to famine and not to assault, and it is still more powerful now. Mayence is on the junction of the Main and the Rhine ; Coblentz and Ehrenbreitstein near the confluence of the Moselle, Rhine, and Lahn ; Strasburg on the junction of the Ill and the Rhine ; Lyons on that of the Rhône and Saône. Every additional stream or river

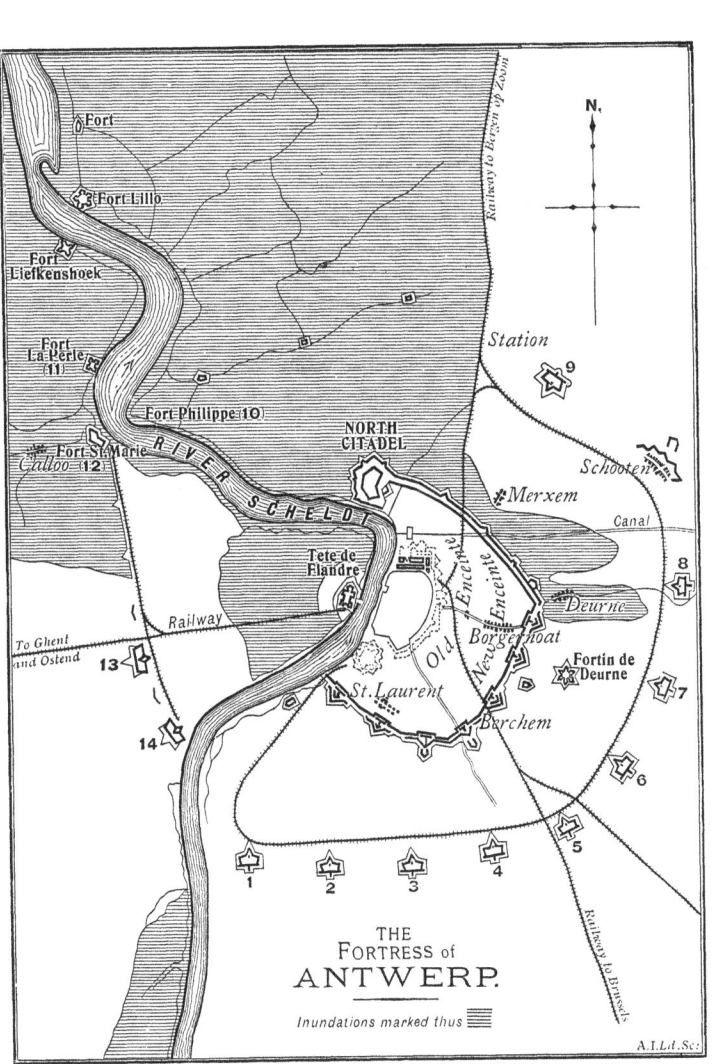

N.

Fort

Fort Lillo

Fort Liefkenshoek

Fort La Perle (11)

Fort Philippe 10

Fort St Marie

Calloo 12

Station

RIVER SCHELDT

NORTH CITADEL

Merxem

Schooten

9

Canal

8

To Ghent and Ostend

Railway

Tete de Flandre

Old Enceinte

New Enceinte

Deurne

Borgerhoat

13

Fortin de Deurne

7

St. Laurent

Berchem

6

14

1   2   3   4

5

Railway to Bergen op Zoom

Railway to Brussels

THE
FORTRESS of
ANTWERP.

Inundations marked thus

A.I.Ld.Sc.

increases the toils of the besiegers when completing their lines of investment. Namur, on a lofty eminence at the junction of the Sambre and Meuse, was a famous fortress in the days of our William III., and to-day, as restored, is one of the bulwarks of Belgium. In addition to these a student will be interested in observing that nearly all the mighty fortresses on which the States of central and western Europe rely as *entrepôts* and pivots for their large armies are on main roads at the passages of rivers, except of course when they are intended as defences for harbours, and in this case also the fact that a river generally flows past their site into the sea increases their security.

Thus Antwerp has long been regarded as an outlying work of England.

In Russia Kovno is on the Niemen, Bobruisk near the Pinsk marshes is on the Beresina, Goniadz on the Bobra. The Polish Quadrilateral consists of Novo Georgievsk on the confluence of the Narew and the Vistula, Warsaw on the Vistula, Ivangorod at the junction of the Wieprz and the Vistula, and Brest Litovsk at the confluence of the Bug and one of its tributaries.

The Austrian fortresses, Cracow and Przemysl, are on the Vistula and the San respectively. In Germany Königsburg is on the Kurische Haf and river Pregel, Dantzig at the mouth of the Vistula, Thorn on the Vistula, Posen on the Warta, Breslau on the Oder, Cologne and Deutsch on the left and right banks of the Rhine, opposite each other. There are several old forts on the Vistula between Thorn and Dantzig, and on the Rhine are also New Brisach and Germersheim, both small places. Hamm is near Düsseldorf on the Rhine.

Various rivers start from the Maloggia block in Switzerland. The sources of the Rhine, Rhône, Reuss, and Aar are quite close to each other, the pass of the Gothard (and the railway tunnel) gives the only entry from Italy into their upper valleys, and this is now protected by many forts. The valley of the Rhône itself is protected by St Maurice, and that of the Rhine by a work at Luziensteig near Sargans. Even if driven from their outer entrenchments the Swiss would find a kind of natural inner entrenchment on the Thur and the Glatt, and could detain an enemy coming from the north or east for a long time on the position of the Limmat with the right pivoted on the Lake of Zurich, and the left at Brugg near the confluents of the Aar, the Limmat, and the Reuss. In this position Massena made a brilliant resistance against the Austrians when the French cause seemed lost in Switzerland, in 1799, but it must be admitted that Massena was dogged and determined to a rare degree.

One of the most celebrated river-fortresses in the world, and one of the best situated, is the present capital of Servia, Belgrade, which is placed at the confluence of the Save, a fine navigable river, and the Danube. As a commanding position in the debatable land between Christendom and Islam it was frequently occupied and reoccupied by both sides. Tremendous battles were fought in its neighbourhood. An island, formed where the two rivers meet, is called the "Island of War." It was one of the depôts of Suleyman the Magnificent, whose military enterprise and knowledge were far in

advance of any European of his time.   The dwellers by
the bank of the Save, perpetual champions of the Christian faith, had special privileges and were organized
especially for war up to 1878.   As the great military
road of the southern Morava led to Belgrade, and as
this was the Ottoman line towards Vienna, the fortress
was named by the Turks " The Gates of the Holy War."
Among the heroes who stood for its possession against
them were John Hunyady in the 15th century and
Prince Eugene in the commencement of the 18th.   It
was not till 1867 that the Turks handed over Belgrade
to Servia.

The old Austrian Quadrilateral in south-east Italy
was an admirable adaptation of the art of defence to
geographical situations.   The course of the Mincio from
the mountains to the Po is short, on it are Peschiera
and Mantua ; the Adige before flowing into Italy gives
an avenue into the heart of the Tyrol.   The road from
Milan to Vienna must needs cross these rivers before
reaching Udine, and having traversed the Col de Tarvis,
the Drave, and the Semmering Pass, attains the old
capital of the Holy Roman Empire and of modern
Austria-Hungary.   It is perfectly obvious that the fortresses of Peschiera and Mantua on the Mincio, and of
Verona and Legnago on the Adige, existing in such a
circumscribed space, absolutely closed all the avenues of
invasion of Austrian territory.   Mantua in 1796 to 1797,
as long as relieving armies frequently appeared on the
scene, tried the energies of Bonaparte to the utmost.
Napoleon III. did not test the Quadrilateral at all ; and
in 1866 the Archduke Albrecht tossed the Italian army
away with contempt.   But the Prussians after their

victory at Sadowa rewarded their Italian allies by com-
pelling the Austrians to evacuate Italy.

The Turkish Quadrilateral before the Treaty of
Berlin in 1878 consisted of Rustchuk and Silistria on
the Danube, and Shumla and Varna. These, with the
double ramparts of the Danube and the Balkans, gave
security to Constantinople as long as Turkey had a field
army ably handled: indeed, the garrison of Silistria commanded by two British officers stopped the Russians in
1854. This was one of the causes which induced the
allies to embark at Varna for Eupatoria in the Crimea,
but these fortresses did not seriously affect the Russian
movements in 1877. Plevna, an improvised fortress in
the hands of Osman Pacha, was of more value, and detained the main Russian force before its detached forts
for nearly five months. The Russians dared not cross the
defiles of the Balkans with 40,000 Turks within striking
distance of their only line of communication, the road
to the Simnitza—Sistova bridge. But once Todleben
completed the investment (November 8th) the fate of
Osman's army was certain. When supplies ran short
he would try a sortie which, like all sorties under such
circumstances, must necessarily fail. He was defeated
with great slaughter, and surrendered December 10, 1877.
Had Osman evacuated Plevna before the investment
was complete he might have renewed the Plevna experiment with even greater success at some other position,
at any rate another field army would have joined Suleiman's between the Balkans and Constantinople.

When dealing with improvised fortifications it may
be as well to refer to remarkable American fortresses
built after war had commenced, and extended and

strengthened under the fire of besiegers as was Sebastopol. Richmond on the James river in Virginia, Vicksburg on the left bank of the Mississippi, in the State of
Mississippi, and Atlanta near the Chattahoochee in
Georgia, are remarkable as objects of striking strategic
enterprises, and defensive and offensive siege tactics
of rare ability and determination. Though in each
case the works sprang into existence with the exigencies of the day, in no case could the besiegers carry
them. Each fortress fell only when it was cut off from
connection with the outer world and no field army was
available for its relief. One of the works of Vicksburg,
Fort Fisher, suffered one of the heaviest bombardments
possible, for Admiral Porter poured into it from his
thirty-three vessels on the Mississippi one hundred and
fifteen shells a minute on December 24th, 1862. Next
day a similar fire was opened, but when the Federal troops
made an assault it was found that all this tremendous
artillery expenditure had been wasted, and the garrison
easily repulsed the assailants. During this siege the
Federals proposed to divert the river into a new bed so
as to leave Vicksburg high and dry, but the Mississippi
was too strong for Grant's engineers and swept away
their work. On July 4th, 1863, Pemberton, the Commandant, who had neither food nor ammunition, surrendered after a defence of 213 days, and, as President
Lincoln said, "the Mississippi ran unvexed to the sea."

Richmond and Petersburg held out from June 15th,
1864, till April 21st, 1865, every assault being repulsed
with terrible slaughter. Grant continually compelled
Lee to stretch his weak line by moving westward. As
he moved his investing line westward so Lee was obliged

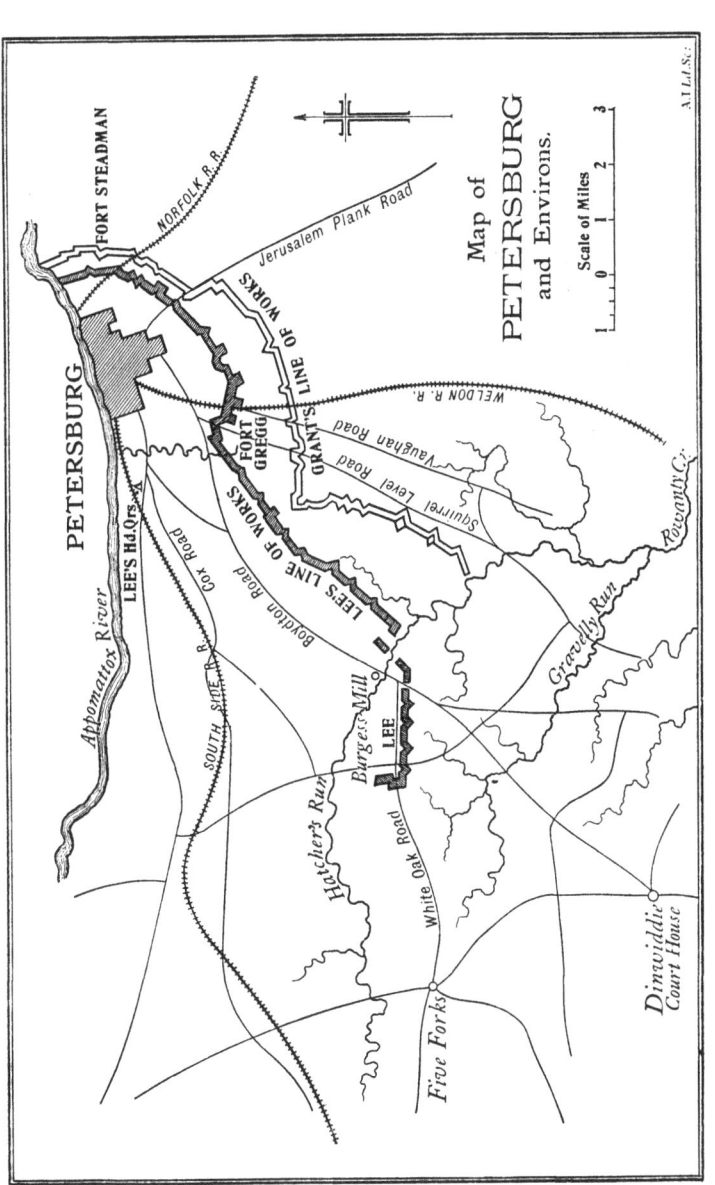

Map of
PETERSBURG
and Environs.

Scale of Miles

to enlarge his works. This could not go on indefinitely, and when Sheridan joined hands with Grant from the Shenandoah Valley, which he had devastated, Lee was in danger of being cut off from every avenue southward. He then abandoned both places and met with the usual fate of generals who try to break through superior investing forces. He was headed by Sheridan and beaten after a fight of three days' duration at Five Forts, being eventually compelled to surrender at Appomattox Court House April 9th, 1865.

Quebec, the capital of the province of the same name, has a population of 62,000, and is thus the third city in Canada. From a strategic point of view it is the most important place in the Dominion, completely controlling the St Lawrence, to which it can admit friendly vessels, and from which it can bar out all hostile fleets. The history of every war fought on Canadian soil shows that the possession of Quebec is essential to the mastery of Canada. The place is described as "the most picturesque and the most strongly fortified city on the Continent." It was formerly a walled city, but several of the old fortifications have been demolished, and some of the gates have been removed. The chief fortification is the Citadel, which stands on Cape Diamond, 333 feet above the river, and covers an area of forty acres. A large factory for the fabrication of small-arm cartridges and artillery projectiles is located at Quebec. The harbour of the city is excellent, and its extensive docks are among the best in the world.

A very interesting study would be to trace the history of the fortresses of the basin of the Seine from the Château Gaillard of Richard I. to the stupendous

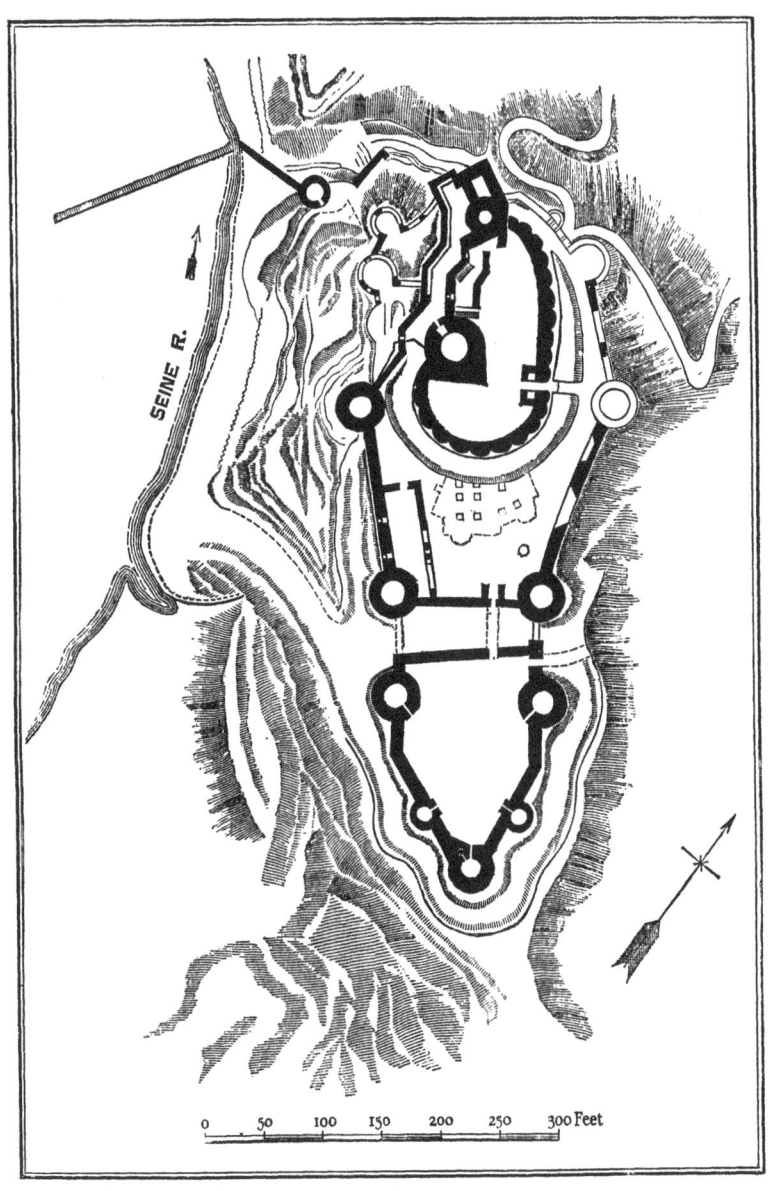

SEINE R.

0   50   100   150   200   250   300 Feet

PLAN OF THE CHÂTEAU GAILLARD.

14

works that now surround Paris. As fortification had reached its highest development under the old system in France in the days of Louis XIV., and as in its newest development of detached forts and enormous entrenched camps it has again reached its maximum in France, it may be well to set forth in a general fashion how that country is now fortified, and how its topographical features might be utilised to give it much greater security at less than half the cost in men and money.

The Archduke Charles, an able writer as well as a brilliant soldier, says, " All that a great country requires is time to develop and organize its resources," and from this point of view great places of arms may be of the greatest utility. A country may be surprised at the beginning of the war and beaten in a few great battles, and if it has no rallying points or centres of supply protected from a rush, and if its capital be open, the enemy might overrun it in all directions with impunity, though it had ample supplies of men and material resources, such as food, arms, and ammunition. Thus Strasburg, though an old-fashioned fortress, resisted Von Werder from August 10th to September 27th, a delay which enabled the French to meet him in force when he arrived on the line Dijon—Besançon. Metz detained fifteen German divisions of infantry and one of independent cavalry from August 19th till October 27th, and thus, as all the other German divisions were around Paris after September 19th, gave the genius of Gambetta full scope to collect together enormous armies under D'Aurelle de Paladines, Faidherbe, Cremer and Garibaldi, which, by the beginning of November, began to show signs of

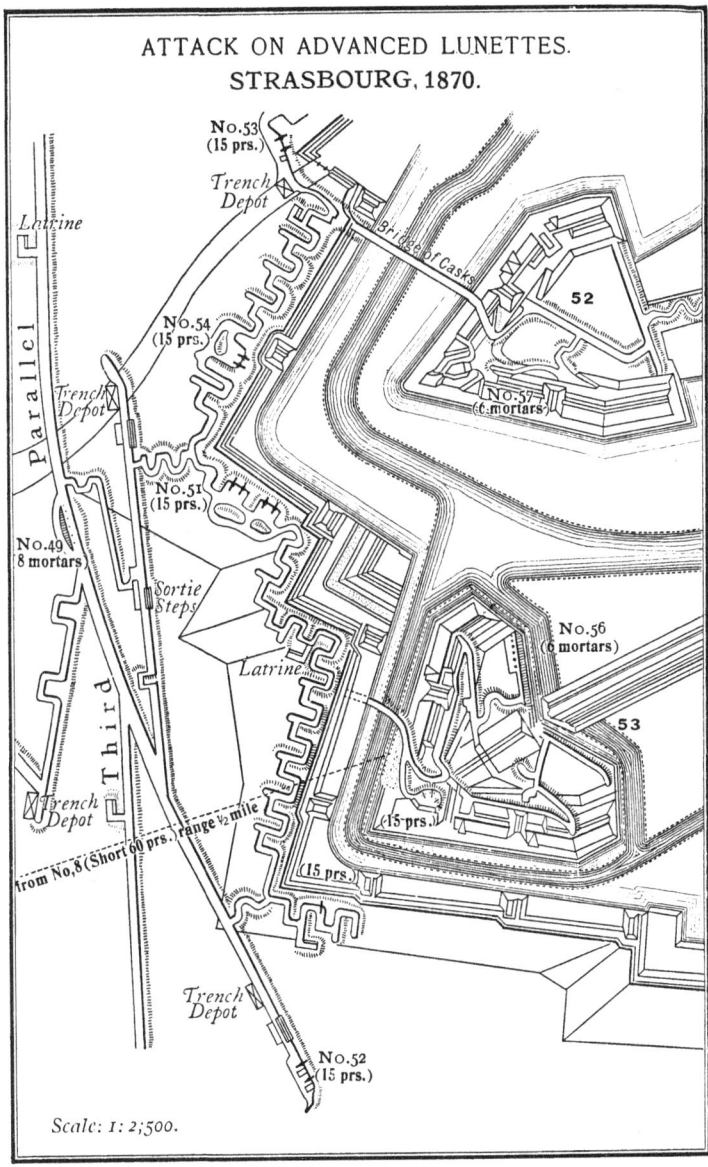

ATTACK ON ADVANCED LUNETTES.
STRASBOURG, 1870.

No.53
(15 prs.)

Trench
Depot

Bridge of Casks

Latrine

Parallel

52

No.54
(15 prs.)

Trench
Depot

No.57
(6 mortars)

No.51
(15 prs.)

No.49
8 mortars

Sortie
Steps

Latrine

No.56
(6 mortars)

Third

53

Trench
Depot

(15 prs.)

from No.8 (Short 60 prs.) range ½ mile

(15 prs.)

(15 prs.)

Trench
Depot

No.52
(15 prs.)

Scale: 1 : 2;500.

dangerous activity on the Doubs, the Somme, and the Loire, and drove the Bavarians out of Orléans and defeated them at Coulmiers. Had Bazaine been able to detain the first and second German armies one fort-night longer round Metz the general opinion is that the corps round Paris would have been in a very critical position, and that their third and fourth armies would have abandoned the siege. But when Bazaine gave up his enormous army and the splendid fortress on the Moselle, the invaders forthwith marched three corps towards the Loire, two between the Seine and the Somme, and left one east of Paris, while another bombarded Mezières and other northern towns. Faidherbe was beaten, D'Aurelle de Paladines was beaten, Orléans was recaptured, Chanzy was beaten, and in the east Bourbaki was driven from Besançon through Pontarlier into Switzerland. Of course, when once the relieving armies from the south and north were driven back, the fate of Paris was that of Richmond—every sortie failed, indeed, none had the smallest chance of success, and the city yielded to starvation January 28th, the bombard-ment, which had been delayed till January for lack of the siege guns, not having produced the slightest effect on the spirits of the defenders.

The present fortifications of France are stupendous, yet unfortunately, in the opinion of the best experts, in many cases useless from a strategical point of view, and certainly antiquated as against modern shells and guns. Numerous works have been constructed with the object of closing railways against the invader, but this is a waste of men and money. The destruction of tunnels, bridges, and rocky cuttings—such as Villers-

Cotterêts, the bridge over the Oise between Creil
and Chantilly, and the bridges and tunnels between
Nanteuil and Paris—delayed the Germans much longer
than the places of Toul, Soissons, La Fère, and Laon.
After the war Germany reduced the number of its
small places to a considerable extent, while France
increased its *forts d'arrêt*. In 1882 there were in France
147 places or independent forts, and 434 out-works and
detached barriers; their mere enumeration occupies
thirteen pages of General Pierron's work. The expense
of construction and maintenance is a terrible drain
on the resources of the State, but the waste of men in
garrisons who ought to be with the armies in the field
is a far more costly charge, and perchance a fatal
injury to the State. In 1880 the *Commission of Defence*
laid it down that the various sections of the fortified
frontiers would require garrisons as follows:—

From Dunkirk to Meubeuge . .	58,663
From Rocroi to the Rhône . .	129,520
From the Rhône to the Col de l'Argentière	26,956
Along the Pyrenees	23,781
On the Ocean Coasts	44,654
On the Mediterranean Coast . .	35,732
In Places of the Second Line . .	36,239
At Paris and Lyons	162,649

New places and new forts have since been constructed and the burden proportionally increased.

If a nation will have forts and places, instead of
placing them in a cordon *along* the frontier a study
of strategic geography would suggest that they should
be placed *perpendicular to* the frontier, just as strategy

suggests when possible a flank instead of a front attack. If the Germans invade again, the line of retreat of the French should be south or, if west, towards Orléans, and not towards Paris. The head of the valley of the Saône could serve as a place of refuge if Gray as well as Dijon, Auxerre, and Besançon were fortified. From these places they could issue—as Bourbaki proposed to do in January, 1871—against the flank of the invader, or against his line of communications by Langres or Épinal. The invader could not live between Gray and the Faucilles; he could not attack from the north, nor could he turn the position from the east because of Belfort and the Jura, nor by the west because of the mountains of the Côte d'Or. It is true that the second and seventh German corps were allowed to traverse these with impunity in bad weather in January, 1871, but such fatuity would scarcely recur. From this line the defenders could issue into the valley of the Rhône by Belfort, into that of the Moselle by Épinal, into the valley of the Meuse or the Marne, into that of the Seine by Dijon, or into the basin of the Loire by Chagny, and provisions would be secured by the railways traversing the valleys of the Saône and Rhône, and Switzerland.

Fortresses to protect the passages of rivers are, as we have seen, quite in accordance with strategical principles, and if the Loire passages were fortified, an excellent line of defence which would defy the invader would be from Orléans to Nevers, and from Nevers by Chagny to Besançon ; this would cover the greater part of France. Provisions and arms could be procured from both Atlantic and Mediterranean ports, railways are numerous, the forests of Orléans and the Nivernais and Morvan would mask manœuvres, and close at hand would be the arms-factories and iron-works of St Étienne and Creusot.

The great danger of entrenched camps, as was fully evident in 1870, is that they prove a fatal attraction to field armies which pivot on them. After any check or temporary disaster the troops fall back into them, get shut up, and surrender. Thus Metz entrapped Bazaine, and even the small fortress of Sédan was injurious to MacMahon; he could not have selected a worse position than to fight with his back to its narrow entrances. But the modern French system of fortified regions, consisting of great fortresses linked together by forts is even worse, and "A.G.," a very able French writer on military subjects, ridicules "The Regions"—the line which extends from Toul to Verdun, from Belfort to Épinal, and from Rheims to La Fère.

When a State proposes to invade a neighbour it ought first to organise a solid basis of operations, a zone from which it can raise reinforcements, provisions, and ammunition. To this zone recruits and stores would come from every part of the interior, and from this zone they would be sent after the actual and active army in the field. On the zone there should be some strong *point d'appui*, otherwise it would be liable to raids and enterprises against its flanks, which would for a time at least destroy its utility, and interfere with the efficiency and continuity of the field army.

An advanced work such as Metz on the new German frontier does not meet these requirements, but the line of the Rhine, Strasburg to Cologne, does admirably, and these fortresses are quite up to the requirements of provident strategy as well as of the most modern theories of fortification.

When the Archduke Charles constructed the works of Ulm, and turned it into a formidable entrenched camp, he little imagined that its advanced position would make it and a large army the prey of Napoleon, and cause the German Empire to afford the astonished world for the first time the curious spectacle of the capitulation of myriads of men *en bloc* without a fight. It is true that Ulm retarded Moreau for six weeks in 1800, but Mack put 80,000 men into it in 1805, hundreds of miles away from his Russian ally. Napoleon from the Main at Wurtzburg and Mayence, and the Rhine between Mayence and Strasburg, closed every road from Ulm to Bohemia, Vienna, and the Tyrol, destroyed or captured every part of Mack's force that tried to break out of the fortress north or south of the Danube, and the unfortunate General, threatened with the direst penalties by his conqueror, surrendered 30,000 men, leaving Napoleon to march unopposed to Vienna.

It would be interesting to dwell on the most famous sieges in European history, and see how far they relate directly to strategic geography, but they would not throw any further light on the subject of the value of fortresses, the dangers of elaborate and far-reaching systems of frontier defences, and of over-multiplication of forts and entrenched camps.

A word or two, however, may be said upon the siege of Paris, September 19th, 1870, till January 28th, 1871, which was certainly the most stupendous operation of the kind ever undertaken. Not to speak of its admirably constructed enceinte, each portion of which was linked to

the other by rail, the line of its outer forts was about
33 English miles long, so that the Germans had an
inner investing line of 50 miles in length, and an outer
line, where the head-quarters of the various corps were
placed, of about 70 miles. For this line they had only
240,000 men, and they were liable at any moment to
have to cope with sorties by the Commandant, Trochu,
and assaults from some of Gambetta's new levies.
Trochu had 400,000 men under arms, of whom at least
150,000 were capable of being made into good soldiers,
and they had ample supplies of weapons. It turned out
that they were possessed of much greater quantities of
food than was supposed, and, strange to say, the most
frivolous population of any city in the world soon
resigned themselves with cheerful fortitude to bear not
only an absolute lack of amusements, but ultimately
the utmost distress for lack of food. To carry the city
by assault was quite impossible, to break through the
forts was equally so. The city was too big for an
ordinary bombardment, and owing to the solidity of the
splendidly built houses the people suffered little from
the fire of even the heaviest guns when these could be
brought up, which was not till January. The bombardment of a given section of the city could not diminish the
resolution.of the other sections, and a park of artillery
sufficiently large for a general bombardment could
neither be brought from Germany nor supplied with
ammunition if on the field. Indeed the resources of the
German commissariat were severely tasked to feed their
soldiers. At last the rations of the French were practically
exhausted, so much so that the immediate reprovision-
ing of Paris was one of the clauses of the Capitulation,

and London alone sent food forthwith to the value of £80,000. The question as to whether the failure of the sorties to break the German investing line of less than 5000 men to a mile was due to lack of genius on the part of Trochu, and caution or cowardice on the part of his officers and men, or whether success was impossible by reason of the impossibility of effective deployment or of flank attacks has been warmly discussed. At any rate the Germans beat back every rush from the gigantic fortress with comparative ease, nor was there any effective strategic coöperation between the vast garrison and the new armies on the Somme and the Loire.

The art of field fortifications as understood by antagonists such as we have to deal with in Asia and Africa, and as applied against them, is interesting, for it illustrates the advantages derived from the most simple defence-works in such wars. Behind their rude stone-works or *sungas* the Ghilzais or Pathans will remain to meet our soldiers at close quarters. The stockades of Cachar, of Perak, and of Burmah, again and again afforded stubborn resistance. The trenches of Tel-el-Kebir on the other hand, constructed on scientific principles, and of formidable profile, prove how serious an obstacle earthworks present to storming columns advancing in compact formation. But it is the defence-works devised by ourselves to meet the exigencies of irregular warfare that are the most significant in pointing the moral. Zaribas—mere enclosures of thorny abattis—proved in the Soudan a sufficient protection against the onslaught of the Arabs. At Rorke's Drift a mere parapet improvised at a moment of desperate emergency out of mealie-bags and biscuit-boxes, enabled the handful of

defenders to keep at bay swarms of Zulus flushed with
the success of Isandhlwana.   Waggon-laagers have
become a recognised mode of defence in South African
warfare.   A simple breastwork sufficed to secure Fort
Battye during the Afghan night attack.   The post of
Dubrai, near Kandahar, protected by a 4½ foot wall,
held out till the ammunition failed.   Haybands proved a
most useful obstacle around Suakin, and mines were
used with good effect.   In the Naga hills a form of
stake called "panjee," consisting of split bamboos barbed
to prevent removal from the ground, was found a serious
impediment in the attacks around Konoma.

A small fortress held to the bitter end may determine
the fate of a State ; and the premature surrender of one
by even two days may ruin a campaign.   In 1566 the
insignificant fortress of Sziget, in Hungary, commanded
by Zringi, detained the enormous force of the Sultan
Solyman the Magnificent for four weeks.   Its small
garrison of 3000 men caused the Turks very considerable
losses ; Solyman himself died before its walls, and thus
an irreparable loss was inflicted upon the Ottoman
Empire.   On the other hand the fortress of Soissons
on the Aisne was surrendered on demand in 1814 to
Bulow and Winzingerode, and they were thus enabled
to unite with Blucher, who would otherwise have been
utterly ruined.   This surrender was fatal to Napoleon's
brilliant strategy.

Every commandant of a fortress therefore should
hold out as long as he can.   He should " eat his boots,"
as Massena said at Genoa in 1800.   On this point
Napoleon issued very clear instructions in 1811, and
again in 1812.   After many other elaborate details, he

said:—" Every commandant should remember that to him is entrusted one of the bulwarks of our empire, or at least one of the *points d'appui* of our army, and one day's delay in regard to its surrender may be of the greatest possible consequence for the defence of the State and the safety of the army. In consequence he should turn a deaf ear to all reports and rumours as to what takes place in other quarters, and tenaciously hold on to his own post."

Fortifications then may be field works, or *fortifications passagères* to strengthen the resisting power of a military position, or a position which is useful either for a fight or for the movements of armies. In some modern wars, especially in the American Civil War, field works were as much a part of the ordinary transactions of armies as the use of rifles and cannons. In Virginia and Georgia they were erected at every big combat. But they are made entirely for the temporary protection of the field army, and cease to have any value when it leaves them either to advance or to retreat. Permanent fortifications are constructed during peace, and as they are equally important whether the army is close to them or at a distance, they must be so strong as to effectively protect both their garrisons—which should be as small as possible —and the stores which the field army may require. But though they are permanent and should facilitate the movements of an army, the latter should only make temporary and occasional use of them. They should be used for movements, and never for a battle : to fight under their cover is to turn them from their true purpose to a false purpose, and to risk instead of to protect the field army. Field works are compatible with mobility ;

works constructed in advance and permanent, if used for
fighting purposes imply immobility. But any army without mobility has no strategic power and is forthwith reduced to the defensive pure and simple, which means
ruin in time.

To use permanent fortresses as if they were *fortifications passagères* is to subordinate strategy to fortification,
whereas fortification should be the handmaid of strategy.
Napoleon called them "oases in the desert." Their
provisions diminish the necessity for long-drawn convoys
—in itself no small benefit. When situated on rivers
they give the offensive army a safe *débouché* across
towards the enemy; if the offensive fail, they cover a
retreat and check the pursuit. But the retreating army
should be content to refit as quickly as possible in the
fortress and then hurry away. If it stay too long it will
be invested, and probably imprisoned unless another
army is at hand for relief. Places such as Thorn, Posen,
or Strasburg may serve as *points d'appui* for the flanks
of an army, whether it advances or retreats, but here
again the army must not be tied to the fortress.

Lord Roberts says in regard to the North-Western
frontier of India:—"I recorded a strong opinion in opposition to the proposals of the Defence Committee, which
were in favour of the construction of a large magazine
at Peshawar, and extensive entrenched works at the
mouth of the Khyber. I pointed out the extreme
danger of a position communication with which could
be cut off, and which could be more or less easily
turned, for it was clear to me that, until we had succeeded in inducing the border tribes to be on friendly
terms with us, and to believe that their interests were

identical with ours, the Peshawar valley would become untenable should any general disturbance take place; and that, instead of entrenchments close to the Khyber Pass, we required a position upon which the garrisons of Peshawar and Nowshera could fall back and await the arrival of reinforcements. For this position I selected a spot on the right bank of the Kabul river, between Khairabad and the Indus; it commanded the passage of the latter river, and could easily be strengthened by defensive works outside the old fort of Attock[1].

At one time there was a plan in vogue of covering the whole frontier with forts just as the French frontier is covered with works, but even in the event of money being available to build these, a sufficient number of men would not be available to man them.

Holland's defensive works are an elaborate system of inundation. In their arrangements for national defence the Dutch have adopted a system similar to that which, until lately, held good in Belgium. Amsterdam is in any future war to be the Antwerp of Holland. So long, it is thought, as the Dutch tricolour floats over the turrets of the capital, the cause of independence will not be lost. It does not necessarily follow from this, as a small minority of Dutchmen recommend, that only the capital would be defended in the event of a conflict with one of the great Powers; but it seems to be generally understood that the defensive operations of the army will be confined to what has been called "the fortress of Holland," a chain of forts circling North and South Holland and Utrecht. Amsterdam may be regarded as the citadel of this immense fortress,

---

[1] *Forty-one Years in India*, Vol. II. p. 406.

the point upon which the defenders would rally for
a final stand, after they had been driven from Yssel
and the Eem-Grebbe positions, the principal towns
on which are strongly garrisoned. The rivers along
all the positions would be held by specially constructed gunboats, mounting heavy guns, and armoured
against the heaviest field ordnance. The "fortress
of Holland" is, even during peace, almost surrounded
by water. The short land side extends from Haarlem
by Utrecht and Gorinchem to Altena, and is called,
somewhat paradoxically at first sight, the "New Dutch
Waterline." Since the retreat of the French under
the King and Turenne in 1672 by reason of the inundations, the Dutch have placed implicit confidence
in their power of flooding an enemy out of their territory.
Het water is Nederlands bondgenoot is no doubt as
true in the 19th as in the 16th century; but until very
recently too little attention was given to the altered
conditions of warfare. A few years ago, Amsterdam
might almost have been occupied before the necessary
arrangements for opening the floodgates were completed. Much has since been done to insure a prompt
and effective inundation. The art of producing floods
is a subject of careful study in Holland. A single act
of carelessness might cause the Dutch to be engulfed by
their own inundation, or convert a province into an
irreclaimable swamp. A general and promiscuous cutting
of the dykes is, for instance, out of the question. The
extent of the flood would be determined by floodgates,
embankments, and canals especially constructed for the
purpose. The inundations would not only be limited
in length and breadth, but in depth. The water,

wherever possible, would be too deep for wading and too shallow for navigation. If it were necessary to adopt one of these alternatives, it is generally considered in Holland that a navigable flood would be more effectual in checking the enemy than one sufficiently shallow for wading. A considerable time must necessarily elapse before a sufficient number of boats and rafts could be procured for the transport of an army, while the Dutch themselves would possess, as in 1672, a considerable number of the so-called *uitleggers* by which the operations of the enemy's craft might be seriously impeded. The ideal inundation for defensive purposes would be about three miles broad and from 1 foot to 18 inches deep, with a soft clayey bottom, intersected by canals and ditches. A flood of this kind, despite its apparent shallowness, is practically impassable on foot, even for individuals, for if they are not stopped by the mud, they would be met by the network of watercourses, which are necessarily invisible. Moreover, the damp cold mists engendered by such an inundation would be but little felt by the acclimatised Dutch soldier, who is better able to endure cold than heat, while they would be exceedingly dangerous to almost any other troops, who would probably suffer more from sickness than from injuries in action. Owing to the varying height of the land, it would be necessary, in order to preserve uniformity of depth, to divide the inundated district into basins by means of transverse dykes, with openings for the in and outflow of water. The inundations in the "fortress of Holland" would be effected simultaneously from the North Sea, the Zuyder Zee, and the various rivers running through

the kingdom from east to west. In the event of a
severe frost every available means would be taken to
break the ice ; but it is admitted that the defensive value
of the floods would in this case be seriously diminished[1].

In 1673 Marshal Luxembourg took advantage of the
ice to invade Holland, but a sudden thaw compelled
him to retire. In 1795 Pichegru's army utilized the
ice to overrun the country, and the fleet in the Texel
was captured by the French Hussars.

No English writer on strategy surpasses Hamley in
clearness, conciseness, and thorough grasp of his subject.
A quotation from his work therefore, though his view on
entrenched camps does not find favour with some other
authors, may fitly close this chapter. "Fortresses, though
without armies they are unavailing, may give to a
country defensive power that counterbalances the cost
of their construction, armament, and equipment, and the
deduction of their garrisons from the active force. And
if, besides being impregnable to open assault, they contain within their defences everything necessary to the
supply of armies, they may be used as temporary bases
or pivots round which an army can operate with vastly
increased power and latitude of manœuvring. Their
value for this purpose will be immensely increased by
forming round them an entrenched camp, that is a continuous line of detached works enclosing space sufficient
for the assembling and manœuvring of an army."

Hamley is also in favour of intermediate lines
between the frontier and the capital, when the distance
between these two is great, and like Marshal Marmont
he would also fortify the capital.

---

[1] See Mr C. L'Estrange in *Navy and Army Illustrated,* September 17th, 1898.

CHAPTER X.

THE PRINCIPAL ROUTES OF INVASION, AND MAIN LINES OF COMMUNICATION.

AN invader who has sufficient force can seldom be prevented from entering mountainous countries whether by the nature of the roads or by fortresses. Thermopylæ proved a trap for the Greeks. Light troops can always be sent round by some side-path or mule-track and they will open up the principal passes, but the tendency of railways and great built roads is to diminish the importance of smaller defiles.

Although few are now used, there are no less than 232 passages on the Franco-Italian frontier, their geographical position being as follows :—

From the Little St Bernard to Mont Genèvre   76
From Mont Genèvre to the Col di Tenda   83
On the littoral   73

If an invader pierced some small passes feebly fortified he would soon turn the principal forts, and the defence would fall to pieces. Alessandria, the principal centre of Italian concentration in the north, could easily be turned from the north by a march on Piacenza.

The same remarks apply to access into the north-west frontier of India. The Hindu Kush is crossed not

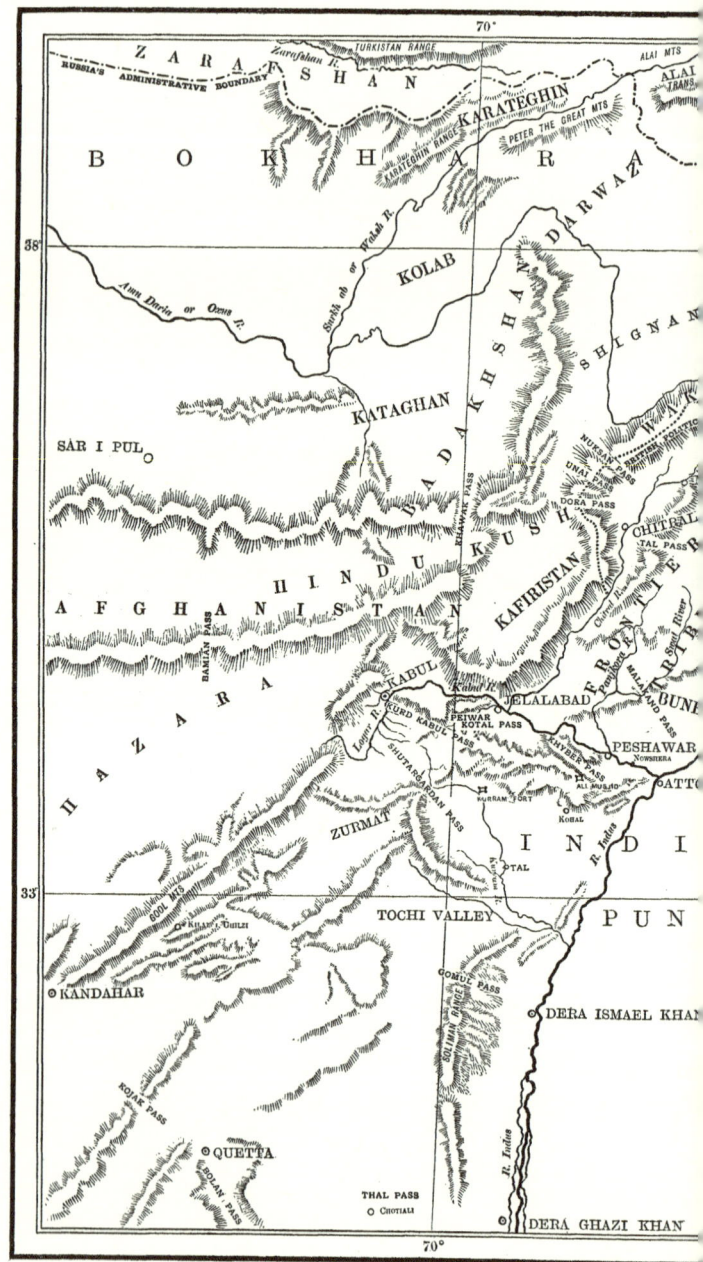

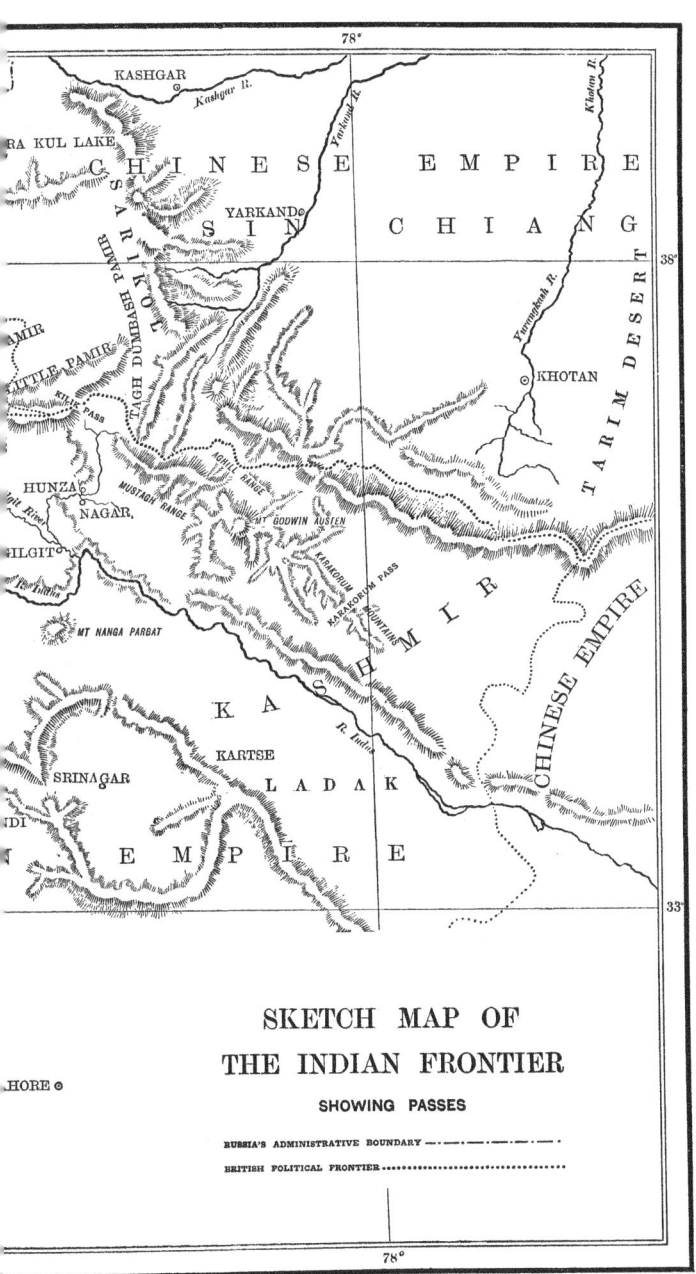

78°

KASHGAR
○ Kashgar R.

RA KUL LAKE

C H I N E S E   E M P I R E

S I N   C H I A N G

YARKAND ○

Yarkand R.

Khotan R.

38°

AMIR

UTTLE PAMIR

KILIK PASS

HUNZA ○
NAGAR

ILGIT ○

MT NANGA PARBAT

SARIKOL

TAGH DUMBASH PAMIR

MUSTAGH RANGE

AGHIL RANGE

MT GODWIN AUSTEN

KARAKORUM RANGE

KARAKORUM PASS

Yurungkash R.

KHOTAN ○

T A R I M   D E S E R T

K A S H M I R

R. Indus

CHINESE EMPIRE

SRINAGAR ○

KARTSE ○

L A D A K

NDI ○

E M P I R E

R. Indus

33°

HORE ○

SKETCH MAP OF
THE INDIAN FRONTIER

SHOWING PASSES

RUSSIA'S ADMINISTRATIVE BOUNDARY —·—·—·—·—·—·—·—·
BRITISH POLITICAL FRONTIER ···

78°

only by the better-known passes as Nuksan, Dora, Alang
and Khawak Irak (Bamian), but by many mountain paths
practicable in summer.   So from Peshawar to Karachi
the hills, as Col. Sir T. Holdich says, are pierced by
gorges and ravines through which excellent roads could
easily be constructed.   The same may be said of the
Paghman range.

The lines of invasion in which the British are more
especially interested are those which lead from central
and north-western Asia into our Indian Empire.

Turning first to the Hindu Kush, we find that Alexander, B.C. 329, passed the ranges S. to N. in 17 days,
and conquered Central Asia as far as the Caspian.   In
327 he passed N. to S. in 10 days.   His main invading
army marched along the valley of the Kabul, while he
led the left wing through the Swat and Buner districts
to the Indus.

The cavalry of the Moghuls crossed these mountains
in myriads at the beginning of the 13th century.   Timurlane's army passed by the Khawak and Tal passes,
September 1398, and Baber frequently traversed the
range from 1519 to 1526.

Passing southward to the Suleiman range, we find
that it has likewise been often penetrated.   Mahmud of
Ghazni led his hordes across it twelve times, 1001—
1027 A.D., by the valley of the Gomul and the Gwalari
Pass to Dera Ismail Khan., Mahomed Ghuri crossed
nine times by the Khyber, Gwalari, and Bolan, 1179—
1195.   The army of Jenghiz Khan crossed by the
Gwalari Pass at the beginning of the 13th century.

Timurlane's right wing advanced *via* Pishin and the
Thal and Chotiali passes to Dera Ghazi Khan, a column

moving through the Gomul valley and the Gwalari. He marched by Shutargardan and Peiwar into the valley of the Kuram, his left wing going by the Kabul valley and the Khyber. The lieutenants of Timur even ventured to cross the barriers separating Samarkand from Kashgaria, which seems at first sight a most desperate enterprise for any army.

Baber in 1527 advanced through the Kabul valley and the Khyber.

Nadir Shah in 1740 advanced from Kandahar to Ghazni and Kabul, and reached the Indus by the Kabul valley.

The British in 1839 and in 1878 advanced on Kabul by the Kabul river valley and the Khyber Pass. The disaster to their army in 1841 occurred in the Khurd-Kabul pass.

Transverse routes connect Kabul and Kandahar, and Kabul and Herat.

It is necessary that we should command the Khyber Pass, and link it with the railway system of India. We should be able to occupy Kabul promptly in the event of a Russian march on India. We must be able to defend the Ameer, if loyal to us, in the last resort with a British army corps at Kabul, connected by railway with our base at Peshawar. This is one great argument for our retaining the command of the Khyber Pass.

The recent Tirah expedition was singular in regard to the utter ignorance on the part of the invaders of the topography of their theatre of operations. The total number of animals employed was 29,470 camels and bullocks, and 42,330 other animals, of which the ponies and donkeys were useless. Mules are far better, and it is false and ruinous economy to employ other animals in mountain warfare. The Jeypore and Gwalior transport trains were far superior to ours.

During this campaign the British penetrated into the recesses of one of the most difficult countries in the world, in regard to which it had been the proud boast of the natives that it had never before been invaded. Their valleys were certainly most rich and prosperous, and their homesteads and farm-yards full of provisions of every description.

A serious difficulty of this country was that it had never been surveyed, and that the scouts had to find out their way as best they could. The course of main streams was merely a matter of conjecture, and the experience of Lockhart's officers gives us a clue as to how barbarians from the Volga and the Danube found their way over the Carpathians and Balkans.

The Sampagha Pass on the south of the Mastura River, and the Arbangha Pass, 7,000 ft. high, on the north of the same river, were much more difficult to climb than any of the passes that any other civilized armies have had to traverse since 1800. The Afridis are the most warlike of mankind. They have good rifles, and they love shooting; they spend their boyhood practising with their rifles, and would give its weight in silver for a good weapon. They had, moreover, some of the best British weapons, and an ample supply of ammunition. Many of them had learned the art of musketry in the British service. So little does he think of the future that the loss of his house, his store of grain, and his fodder is as nothing to the young Afridi, when compared with the present delight of plunder and murder.

He is a master of guerrilla warfare, as fleet of foot as a goat, and can live for days on the grain he carries with him. No necessity, therefore, exists for them to cover a line of communications. This extraordinary mobility enables them to attack from any direction quite unexpectedly, and to disperse and disappear as rapidly as they come. As Captain Shadwell says, when led by Ghazis "or when excited by a fanatical preacher to expect a safe conduct to Paradise in case of death, their daring courage is of the most reckless description." Of all the Afridis, the Zaka Khels were the most desperate and troublesome fighters, and the last to submit; they cut telegraph wires, were persistent in "sniping," attacked convoys, and fired at long range on foraging parties.

Suppose any future treachery on the part of the Afghans; it is necessary that Russia should know, and that the Afghans should know, that we have the power of enforcing loyalty to the solemn engagement made between the latter and ourselves. The operations of General Low and Colonel Kelly for the relief of Chitral in 1895 were most instructive and completely successful, while the defence of Chitral itself was a brilliant affair. The effective of the garrison were only 543, and they stood a siege of seven weeks, and made some good and useful sorties. The dwellers in the mountainous territory are amongst the most martial and daring hillsmen in Asia ; they are desperate Mussulman fanatics in addition to being hardy sportsmen, good at polo and all manly exercises.

The defiles were most dangerous—in fact a British detachment under Capt. Ross was cut to pieces at

The Upper Swat Valley.

Karagh near Buni. Nothing daunted, Kelly marched
to Ghizar by the Shundar Pass, 12,000 feet high, and
hastened to the beleaguered town, while Sir R. Low with
14,000 men advanced from Peshawar, forcing the Mala-
kand Pass against the intrepid Swatis at an altitude of
3,500 feet. As these were retiring, they were charged
by the 11th Bengal Lancers and the Guide Cavalry.
Wellington said he did not use his numerous and
splendid cavalry at Vitoria because of the nature of the
ground, but had Low thought of a *terrain* for cavalry,
he would never have defeated the Swatis; indeed the
doctrine of suitable ground for cavalry would have
prevented the use of this arm in the campaign, for no
country could be worse. If the squadrons at the Mala-
kand Pass had adhered to text-book theories as to their
rôle they would never have wielded either sword or
sabre; in fact, as a French author says, "they could not
have reached a place fit for the theoretical cavalry
manœuvres between the Indus and Siberia." The Swat
and the Panjora rivers were held in force by the enemy,
but they were driven back and the latter river passed
by a bridge of boats. Miankhalai and Dir were occupied,
and thence the march to Chitral was easy. But the
enemy, weary of their vain efforts against the little
fortress and aware of the advance of Kelly and Low,
disappeared.

When the Pamir plateau was partitioned between
England and Russia, the latter country was allowed
in that partition to get somewhat inconveniently near the
extreme north-western corner of our frontier. Chitral
lies at the base, and in the event of war or trouble with
Russia, her scouts would cross her Pamir frontier, and

traversing our piece—for we never keep any establishments in that snowy region—they would at once discover
and report what was going on in the upper valleys of
the Indus.   Now with these movements unchecked, the
moral effect on Kashmir and the outlying portions of
the dominion would, without doubt, be very bad indeed.
Sir Richard Temple says there is a danger lest our

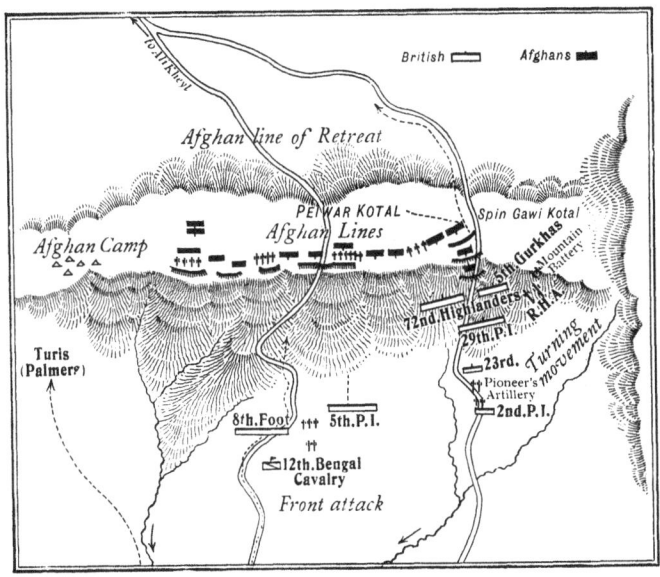

ATTACK ON PEIWAR KOTAL, DECEMBER 2, 1878.

holding of the Khyber Pass should set up the Afridis or
keep the war, like fire, still warm.   If this has been
decided upon, we should spare no pains to assure the
Afridis, in common with all the tribes, that their independence under us is absolutely secure.

Near Bannu lies the Kuram valley, and from this
valley there are two passes which lead direct to Kabul.

As evidence of the advantage of this route by the Peiwar Kotal Pass, which is now the British boundary, and the Shutargardan, or Camel Neck Pass, which leads straight down to Kabul, Sir R. Temple suggests that opposite Dera Ismail Khan, and opposite the Gomul Pass, a viaduct ought to be constructed, so that we may be able to take an army corps across the river at will, if necessary, because it would be a second string to our bow in the event of anything happening at Kandahar. This, indeed, is a matter of first-rate military importance—how important we should never really know until something had gone wrong with us at Kandahar.

There is a great railway viaduct at Sukkur some 250 miles to the south. Skobeleff is reported to have said that, "not for a generation or more to come would Russia be able to advance beyond Herat upon India. But in the meantime by the railway from Merv we are assuming a menacing position towards England, which will keep her occupied in India and prevent her impeding us in other parts of the world," to wit Constantinople. These words are pregnant with meaning. The railways of Russia are, of course, largely extending. There is first of all the Trans-Caspian, which runs from the eastern shores of the Caspian Sea to within striking distance of the Herat frontier. But the Russians propose to make a railway from Merv also, which comes to the same thing practically.

Then from Merv the railway has for some years past extended to Bokhara and Samarkand, and now from Samarkand it has been completed to Tashkend, with a view to an ultimate connection with Balkh, on the Oxus river. Russia has now got her railway extended from the main trunk line of Siberia, which runs from west to east, and from that trunk line she connects with Orenburg, which is the south-west corner of Siberia. From Orenburg, in the course of a very few years, she will have a railway on to the Sea of Aral, and thence to Khiva, and will thus complete a regular circle by the Trans-Caspian railway to the Caspian Sea.

There are two lines of possible Russian advance to India. One would possibly be from Samarkand or Bokhara to the north of the Oxus near Balkh, on to the Indian Caucasus by the valley of the Oxus to Bamian, and so on to Kabul. This is the first line, and this is the hardest and least probable. The second is from Baku, across the Caspian Sea to Michailovsk, the terminus of the railway on the eastern shore of that sea, then on to Merv, and so to Herat and Kandahar.

Sir R. Temple declares that it is sufficient to hold up one's little finger and a money-bag in any part of India to be able to raise any number of cattle and camels, but recently this phase of the transport question has been attended with difficulties. It is not, however of great importance, because we have railways on to our fighting point. On the other hand, at their southern line the Russians have a march of 370 miles, and at the upper of 700 miles, over a very difficult and desert country.

Inside the mountains Quetta has become a great basis of military operations, which would be of supreme importance in the event of war with Russia. From Quetta there runs a line through Pishin, piercing the Khoja Amram range by the Khoja tunnel. For military strategy, in the event of extreme danger, there is no more important work in the British Empire than this tunnel, which takes us straight on to the plains that lead to Kandahar. The line is now complete as far as New Chaman, leaving 70 miles to Kandahar to be completed.

If Russia attacks India by the south-western route, which is the most likely one, it is at Kandahar that we should give her battle. In every way this is a valuable military position. The commander of a British army-corps occupying it would have a desert on his left, which could not be turned; he would have a river on his right, which could be turned only with great difficulty; and in his rear would be one of the richest districts on the face of the earth, limited no doubt in area, but boundless in fertility. He could draw his supplies from this, and would have a temporary railway—which could be made in two and a half or three months at the rate of a mile a day—running from Kandahar to Chaman; whence, of course, there is railway communication with India. In front a rough and half-desert road of 300 miles or more is the only means of approach for an enemy. That this is the most advantageous position for the British from the military point of view there can be little doubt. The political advantage, of course, is enormous, because it is in the highest degree inadvisable that India should see us fighting with Europeans. Over that tremendous transaction it is best that there should be a veil, to use Sir R. Temple's words.

Russia is building a railway as fast as she possibly can from Merv to Koshat on the Afghan frontier, seventy miles from Herat. It may be possible to explain the Trans-Caspian railway to Merv and thence to

Bokhara, Samarkand, and Tashkend as commercial, but the Merv-Afghan line is purely strategic, and is a menace to our Empire. The Russians are also examining our frontier near Siestan, and if they could secure as powerful an influence in southern Persia as they have acquired in the north of the same state, the position of affairs would be still more serious.

Our own frontiers now march with the Pamirs, and our position on the borders of this wild region is not unsatisfactory at the present time. Before Colonel Durand's successful campaign in 1891 against Hunza and Nagar, Russian officers with a few Cossacks had visited Hunza and also Chitral, unsettling by their presence the neighbouring tribes. After 1891, the question of the garrisoning of the Gilgit Agency and the connecting of that province with India was taken up. The garrisoning was entrusted for the most part to the carefully drilled Imperial Service troops of the Kashmir Government, with a considerable number of British officers attached to inspect and train them, and take command in the event of hostilities. The present garrison of the Gilgit Agency, which borders the Pamirs, consists of two companies of the 42nd Gurkhas, two Kashmir Imperial Service infantry battalions, and one Imperial Service mountain battery, with a few sappers—some 2,000 men in all. There are outposts at Gupis, *en route* to Yassin and the Baroghil Pass leading to the Pamirs, also at Hunza, near which the Kilik Pass also leads to the Pamirs. The communications between our Agency and Kashmir and thence with British India are closed by several lofty snow-bound passes, which are only open for some six months in the year   During these six months the Indian Commissariat puts in supplies for the whole year for the garrison of the Agency.

A telegraph line, with difficulty maintained in winter, spans the 200 odd miles which separate Gilgit from Kashmir, while excellent roads for pack traffic now exist along the main line of the Agency. The force of our troops therefore—for the Imperial Service troops are as much our troops in a broad sense as any other[1]— is ample to prevent any attempts Russian officers may make at encroachment, or as they prefer to call it, scientific research, so far as our own territory is concerned, whilst encroachment on Chinese ground can only be resisted by diplomatic means.

At one time alarming views were held as to the possibilities of Russia's threatening India *viâ* the Pamirs, but happily further acquaintance with the country bordering on the Hindu Kush has dispelled these. Enormous passes from 13,000 to 16,000 feet above the sea, mere goat-tracks, snow-bound for three parts of the year, separate Russian outposts from Gilgit and Chitral, while many passes almost as lofty separate these districts from India and Kashmir. The principal passes of the Pamirs are, from east to west, the Kilik, the Baroghil, and the Dora, leading into Hunza, Yassin and Chitral; between Gilgit and Kashmir are the Burzil, 13,000 feet, and Tragdal Passes, 11,500, as well as minor ones ; between Chitral and India, the Lowarai Pass, close on 10,000 feet. Forage and transport are almost unattainable, so that there is little to be feared in that direction should Russia and India come to

---

[1] *The Broad Arrow*, May 6th, 1899.

blows. Prior to the Hunza and Chitral campaigns, matters were no doubt different; small parties of Russians could, and on .two occasions even did penetrate far enough in to raise many a rumour of invasion which would grow and increase as it travelled south. The Russian officers who met the British party in 1895 to settle the Pamir boundaries, openly confessed that we " had slammed the door in their faces." The harm that Russia can now do to us is to oust our trade in Chinese Turkestan, and this she is doing fast. Her railway extensions bring her nearer the trade-area every year, and she will probably ere long permanently supplant all our trade there.

Some of the most interesting enterprises of modern times have followed the planting of British rule in Upper Burma as far as the borders of Eastern Tibet. Dacoity has been suppressed in the woods and fastnesses of the Chingwin and the Upper Irrawaddy, and security now prevails to such an extent as to justify the investment of capital in a railway from Mandalay to the Salween through one of the most difficult countries in the world. Too much credit cannot be given to engineers like Mr Wagstaff, and officers like Capt. Macquoid of the Hyderabad contingent, who have been pioneers of civilization and wealth in these rugged regions amidst semi-savage populations. The description of parts of this territory in the *Journal of the Society of Engineers* by Mr Wagstaff deserves careful study[1].

When Upper Burma was annexed the railway

---

[1] Lecture of March 6th, 1899.

problems that faced the new executive in charge of the country were—

(1) The extension of the existing railway to the capital of the newly annexed province, Mandalay.

(2) The construction of a line north from Mandalay to the upper reaches of the Irrawaddy, to the frontier near or above Bhamo, and

(3) The construction of a line from the valley of the Irrawaddy to that of the Salween, to be eventually extended into the Chinese province of Yunnan.

The first two of these problems have now been solved, and the third is in hand. The route selected for this line, after much reconnaissance and discussion, is as follows :—Mandalay, Gokteik Gorge, Lashio, Salween River, distance 230 miles. The principal difficulties to be faced were, the Ghât between the 15th and 35th mile, with a rise of about 3200 feet, most of which was "bunched" in a shorter distance ; and the Gokteik Gorge, with a fall and rise of about 1400 feet in 8 miles.

In the autumn of 1892, a detailed survey was decided on, and three parties went out ; one to start work on the Ghât, one under the charge of Mr Wagstaff to begin with the Gokteik Gorge, and one to do the work in the fairly easy country between. Mr Wagstaff's party consisted of himself, one senior and two junior assistant engineers, and two native levellers, 50 chainmen and tent khalassies, a clerk, three interpreters, an escort of 15 military police under a native officer, a native hospital assistant in medical charge, a native in charge of the rations, and about 12 or 16 servants and followers. Rations and stores, with the exception of rice, for all natives for three months, were carried, and a medical outfit and tents for the whole party. It was possible to procure rice locally, but towards the end of the season great difficulty was experienced in obtaining the necessary quantity. A further supply for three months was sent up later.

Transport consisted of about 150 pack-bullocks, with a Shan driver to every four or six animals, the load for each bullock being about 150 to 180 lb. These materially increased the size of the camp. The trained Shan pack-bullock, the principal transport animal used in the Burmo-Chinese trade, can travel from six to seven hours a day at the rate of about two miles an hour, if doing continuous marches for many days in succession. For single marches, with a day's rest before and after, he can cover from 20 to 25 miles. In the case of heavy loads that could be divided, Mr Wagstaff adopted the expedient of selecting two specially strong animals, and used them to carry the load on alternate days, one bullock always doing his march without a load at all. All goods have to be made into loads weighing from 75 to 80 lb., and are packed in bamboo crates and slung over a pair of tightly-stuffed cushions on either side of the animal. A neck and a tail rope are used to prevent the load slipping forward or backward when ascending and descending, but no girths are employed. The drivers are very clever at crating and loading goods, and it is best to leave this to them.

Kerosine-oil cases, *i.e.* the wooden boxes that come from America, each containing two tins of oil, if clamped with iron and fitted with lids and hinges and a piece of painted canvas, rather larger than the lid, nailed on to the top as a protection from the wet, form excellent

M.   16

loads. A bullock carries two of these, fitted with stores and a light roll of bedding over the top, very comfortably. A pack-bullock takes about two years to train, and is worth when trained from 70 to 100 rupees, or say £5 to £6.

A curious feature of the hill-sides in the part of Burma under discussion, and a common one on both sides of the Gokteik Gorge, is the existence of large tracts of swampy ground where the swamps are on the tops of the hills and not in the valleys. The ground appears to consist of a continuous layer of marl and vegetable fibrous matter combined, about two or four feet thick, spread over the surface of non-permeable rock. This coating appears to have the power of retaining a considerable amount of water and to need to be hung to dry before it will part with it. On the steep slopes of the hill-side it is thus hung to some extent and dries on the surface, the water running off down the gullies on the bare rock; but on the rounded tops of the hills where the slope is small it remains lodged all the year round and fills every pocket in the rock below with silt and water. In surveying on this ground it is necessary to walk for days perhaps up to the knees in the swamp seeking a solid footing below for the greater part of the time, varying matters by an occasional plunge into a hole of unknown depth. It is practically impossible to use a theodolite, as the least movement shakes the surface for fifty yards round. All work has therefore to be done with the compass and the levels taken round the edge. It might be thought that these swamps would prove formidable obstacles to construction, but experience has shown that this is not the

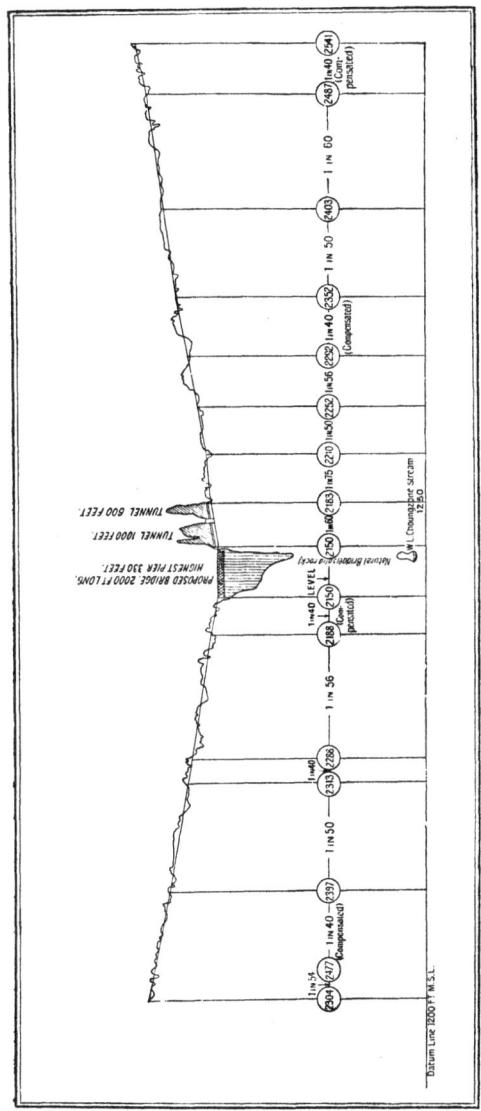

SECTION OF RAILWAY OVER THE GOKTEIK GORGE.

case. A few cross-cuts or drains a few feet wide across the surface of the swamp taken down to bed rock, with pipes or small culverts to carry the water thus collected through the formation, is all that is needed to induce the swamp to drain and give a surface for the bank. In cuttings a large side drain to carry off the collected water is necessary.

The question of malaria was a very serious one for the working parties, both on survey and construction. No working survey-party should go into the Terai or foot-hills until the 1st of December at earliest, and should return if possible by the beginning of May.

The viaduct over the gorge will be one of the largest in the world, measuring over 2,000 feet, and standing 330 feet high. It will require 5,000 tons of steel. It is noteworthy that the contract, in consequence of the inferior tenders in every respect of British firms, has been assigned to Philadelphia.

Two important pieces of exploration have just been made between India and China. One of these is the discovery by Lieutenant Watts-Jones of a feasible railway route from Yung-chang-fu, *via* Monkyeng, to Yin-chow. Hitherto the prevalent idea has been that the mountains between Tali-fu, one of the chief marts in Yunnan, and the British frontier constituted a practically insuperable physical barrier to railway connection between British Burma and Western China; therefore, if a practicable route exists capable of extension to the banks of the Upper Yang-tsze, the fact is one of obvious moment. Another notable journey, made in the reverse direction from Shanghai, *via* Hu-nan, to Bhamo, in Burma, has been achieved by Captain Wingate,

of the 14th Bengal Lancers. The two British expeditions appear to have met in south-western Yunnan,
and no doubt were able usefully to compare notes.
Yunnan and Kwei-chow are described by Captain
Wingate as poor and sparsely populated—a fact, of
course, already well known—but it is interesting to
learn that he was greatly impressed by the wealth and
enormous possibilities of Hu-nan. The same conviction
forced itself in 1898 on Mr M. O'Sullivan, who paid
two visits to Hu-nan, and was astonished in that re-
puted hot-bed of anti-foreign feeling to find the electric
light installed in a number of houses and shops in
Chang-sha. Mr O'Sullivan did good service in endeavouring to pave the way for an official British
mission to this important province, which Mr F. Bourne,
of the Consular Service, has also strongly advised
should be opened up to trade. If the coming railway
is to be pushed on from the Indian side, and if this
be seconded by a friendly mission to the Hu-nanese
authorities from the Yang-tsze, it will mean a most
useful consolidation of our influence in Central and
Western China.

The works executed on the Great Indian Peninsular
Railway, to enable it to cross the Ghâts, are in no respect
inferior to the famous stairs of Giovo or Semmering
between Vienna and Laibach. The total height that
had to be surmounted was 1820 feet, on a line of 15
miles and with mean incline of 1 in 48. It was necessary to construct eight viaducts of from 30 to 50 arches,
and from 50 to 140 feet high, to cut 22 tunnels of a total
length of a mile and three-quarters, and to make embankments containing upwards of 6,000,000 cubic feet.

But the British have also constructed 33,000 miles of excellent roads, worthy of the Romans. They are carried right over mountain-ranges like the Ghâts and the Vindhyas and through difficult passes. The road between the rivers Jhelum and Indus consists of continuous cuttings and embankments for 150 miles. The fact that the old historic dynasties of India utterly neglected roads is the more curious, as manifestly there had been good engineers and good work done on roads by dynasties which have left no historic records. British officers found many remains of old roads north of the Indus which have facilitated their labours in recent campaigns.

In other continents we are also face to face with the strategic consequences of lines of railways, but these are at present complicated in but a very slight degree by questions of fortresses.

It must be remembered that in Europe, as will be seen, excellent roads have existed for centuries close to the lines selected by railway engineers during the past fifty years, and indeed a fair supply of canals opened up means for the conveyance of bulky articles from the sea to the centre of the various countries. But in the greater part of America, Africa, and northern Asia, railways came first, and the lines which connect distant centres of civilization and trade, such as New York, Chicago and San Francisco, were made before the roads; most of the latter linking the railway stations with more recent agricultural, mining, or manufacturing depôts.

The great strategic and commercial line through the Dominion of Canada is of the first importance: it was a daring and beneficent scheme in its conception, the support of which puts the reputation for foresight of the Dominion Parliament far above that of the mother of Parliaments or any other similar institution in Europe.

The topographical details of the regions traversed by the Canadian-Pacific Railway are simple, and may be described as four main regions—the accentuated territory between Halifax and Montreal, the boundary of the Great Lakes, the vast prairies between the Lakes and the Rocky Mountains, and the mountains and rivers of British Columbia. The line links the Atlantic to the Pacific, and strengthens the offensive and defensive of our Empire to a remarkable degree. Troops from China could be in Great Britain by this route in 25 days, while troops from Liverpool could reinforce our naval depôt at Esquimault in Vancouver's Island in 15 days. With regard to the chance of the line becoming blocked by snow, Mr Parkin says that not one day was missed in five years ; though during the same period United States lines were frequently rendered useless for days at a time.

In Canada the central decisive point is Montreal (situated on an island of the same name), having a population of 140,000 and covering an area of eight square miles. The St Lawrence is here crossed by the celebrated Victoria Bridge, an iron tubular structure nearly two miles long, supported on twenty-four piers of solid masonry. As a railroad centre, the head of un-impeded ocean traffic, the outlet of the Canadian system of canals—in brief, as the connecting link between the ocean and the lakes—Montreal is a point of immense commercial and strategical value, and has been termed the "key and the capital of Canada." In strategic importance it is second to Quebec alone.

With regard to the recently occupied region of Klondyke the sufferings of the first settlers on their way to the gold-producing areas were intense. Travelling from the coast to Dawson City was a more risky pilgrimage than the passage of the Great St Bernard by Napoleon. But roads and railways are now being rapidly opened up, and in a few years the terrors of the White and Chilcoot Passes will be things of the past.

We have already seen that the one weak point of the Canadian-Pacific line is the portion that traverses the American State of Maine, but the Inter-Colonial line remedies this defect, and it can be easily defended. Its Pacific section is secure. No American general would care to operate from the Californian base and the mountainous frontier on the west through the gorges and over the torrents of British Columbia. Compared with this an Afghan campaign would be a trifling operation. But what of the district between the Rocky Mountains and Lake Superior?

Raids on the railways on each side of the 49th parallel of latitude—which is the boundary westward from the Lake of the Woods to New Westminster—would certainly interrupt communications, and abundance of supplies could be found by the raiders. Moreover, there would not be any considerable physical obstacles to the flying columns. But no raid in Manitoba, Assiniboia, Minnesota, or Dakota could seriously affect the social life or prosperity of the great coast towns on either side. As long as Great Britain has command of the seas, and especially if she keeps her contract number of vessels in the Great Lakes and holds the canals of communication between them and the St Lawrence, no

American assault in Canada has a chance of success.
It is true the St Lawrence and many other Canadian
waters are frozen in winter, but, in winter, movements of
troops in Canada—which does not afford half the cantonments that are found by the banks of the Lisaine
and Doubs—would entail sufferings as awful as those
undergone by the XVIII, XX, and XXIV French
corps, from January 15th to February 2nd, 1871.   Nothing but some civil strife between Canada and Great
Britain, or some crisis with other Powers which would
shake the United kingdom to pieces, or loss of command of the sea by a defeat in European waters, or a
startling and excessive development of United States
sea power, could imperil either Halifax or Quebec, or
Montreal, or Esquimault, or do more than inflict temporary and easily repaired damage on the railway line
which enables them to co-ordinate their energies.

Perhaps the most romantic and most interesting of any
line of communication in the world is that whereby it is
proposed to connect the extremity of South Africa with
the Mediterranean.   From the point of view of a line for
opening up new territories, rendering available exhaustless stores of all kinds of necessaries and luxuries, and
facilitating military operations in lands which the soldiers
of the Pharaohs, the Ptolemies, or the Cæsars dared not
approach, the railway from Cape Town to Alexandria is
one of the greatest conceptions of modern days.   It is
already distinctly traced and, although more than half
its course is marked unexplored in every map published
forty years ago, it has been largely completed.   It has
conveyed soldiers from regions further away than the
homes of those Goths whose prowess amazed Hypatia,

to the tombs of Alexander and the Mamelukes, and to the junction of the Atbara with the Nile itself, whence the river was ascended to the junction of the Bahr-el-Ghazal, through the *sudd* which bars the progress of navigation to the equatorial lakes. It has also linked Cape Town with the diamond mines of Kimberley, and these with the capitals of Khama and Lo Bengulu, and it is now nearing the Zambesi, and steadily approaching the junction of the Blue with the White Nile. This, if not spoiled by party spirit or paltry economy, will be a monument to the genius of our race more enduring and infinitely more far-reaching in its consequences than that splendid paved Watling Street by which the Roman legions marched across Britain from the banks of the Thames to North Wales.

At the date of writing these lines the route is detailed as follows :—

Rail—Cairo to Assuan, 583 miles ; boat—Assuan to Wady Halfa, 200 miles ; rail—Wady Halfa to Khartum, 560 miles (railway is to be constructed) ; boat—Khartum to Fashoda, 450 miles ; boat—Fashoda to Albert Nyanza, 750 miles ; land journey—Albert Nyanza to Lake Tanganyika, 450 miles ; boat to end of L. Tanganyika, 400 miles ; projected railway to Buluwayo, 860 miles; whence railway opened to Cape Town, 1,360 miles.

A correspondent of the *Times* thus describes a part of the route in September, 1898 :—

" The line gradually ascends from Wady Halfa to a point about 103 miles out, where it attains an altitude of 2,100 ft.—that is 500 ft. above the Nile at Wady Halfa ; from this point there is a gradual dip towards the Nile at Abu Hamed. The railway throughout traverses one of the most utterly desert regions on the face of the earth—flat wastes of yellow sand, here and there ribbed by ranges of bare black crags. In a few spots are clumps of withered camel-thorn, but as a rule not a sign of life, animal or vegetable, is visible ; in every direction spread out under the cloudless blue sky are the glaring sands and the dark rocks, with only the lakes and seas of the never-failing mirage to relieve the monotony of the dreary scenery. I went with an escort of Ababdeh Arabs to Wady Halfa by the desert route, which runs almost parallel to this railway and through a similar country ; but there I saw the more or less well-defined tracks of caravans, the bones of men and camels who had perished on the way, whereas the railway has been carried across a portion of the Nubian Desert which has apparently never before been trodden by the feet of man. In one place, however, it crosses what seems to have been the route taken a long time ago by a large body of men ; for here, 60 miles from the nearest water, the engineers came across a mysterious collection of many hundreds of broken 'zeers' or earthen water-coolers ; they were two-handled graceful amphoræ of a shape unknown in modern Egypt. This discovery naturally started a good deal of conjecture. Some re-membered that Cambyses sent an army across this desert ; then, to come down to more modern times, Ismail Pasha once despatched a force of 2,000 men to Abu Hamed, which entirely disappeared in the desert, and was never heard of again. Said Pasha, too, travelled in luxurious state to Abu Hamed, in a carriage drawn by eight camels, with an army watering the desert in front of him to keep down the dust."

The Sirdar was able, by the splendid skill of Lieutenant Girouard, to construct his railway at the rate of
two miles a day across a country where there were no
tunnels, hardly any cuttings, and no embankments. The
difficulties in the way of the Uganda line however were
considerable, and sometimes unexpected. They are best
described in Lord Salisbury's words :—"A little further
south we are constructing the Uganda Railway across
some 550 miles of unknown country. We have there
cuttings and embankments to undertake. We have no
great command of labour and a limited supply of money,
and we do not go there quite so fast as the Sirdar was
able to go with his Korosko and Khartum Railway.
But still it is advancing steadily, with such accidents as
in such a country might perhaps be expected. We
suddenly learned that we had altogether a wrong notion
of the configuration of the country, and by altering the
line we were able to save 100 miles of our journey. But
there were other surprises that awaited the construction
of the railway in that country. The whole of the works
were put a stop to for three weeks because a party of
man-eating lions appeared in the locality. At last the
labourers entirely declined to work unless they were
guarded by iron intrenchments. Of course, it is very
difficult to work a railway under these terms, and until
we found enthusiastic sportsmen to get rid of these man-
eating lions our enterprise has been seriously hindered.
There are many difficulties ; no water, no food, and a
great disinclination on the part of any of the natives to
work for any consideration whatever. Yet, in spite of
all these difficulties, we have completed more than half
the railway to the lake, and by the end of next year we

shall have reached the lake. Well, that means the sub-
jugation, and, therefore, the civilization of a vast country.
Nothing but that railway could give us the grip over
the country that would enable us to undertake the
responsibilities of so vast a territory."

Owing to the railway the question of the supply of
horses and camels for our expeditions in the valley of the
Nile is not nearly so serious as it once was, but it has
never been solved in a satisfactory manner. For the
Egyptian expedition 5,400 horses were embarked, a
number considerably less than half the establishment of
any army corps as laid down. The three regiments of
cavalry of the line were given an establishment of 465
troop horses in place of 524, and yet required the
transfer of 591 horses from other regiments to enable
them to take the field. The artillery at home was de-
nuded of 934 serviceable horses to bring the batteries
detailed for service up to war strength and to supply
regimental transport. The engineers, with a number far
short of that prescribed by regulation, actually embarked only 10 horses short of the total peace establishment of the corps. Of the horses remaining in England,
2,450 were disqualified by age, and therefore unfit for
service. Thus to place on a war footing 4 cavalry
regiments, 15 batteries, 6 ammunition columns, and full
engineer complement for the establishment of an army
corps, the mounted branches of the service were reduced
to a state of complete inefficiency.

The picture is not encouraging, for in 1878, at a
period of great emergency, it required four weeks to
purchase 2,250 horses, and in 1882 they were bought
only at the rate of 100 a week. One lesson to be learnt

from the only campaign in which of recent years a considerable force of cavalry and artillery of the home
establishment has taken part, is that the condition of
the mounted branches of the service as regards horses
will not bear the strain of the most partial mobilization,
and that the formation of some reserve whence remounts
can be drawn on emergency is one of the pressing
necessities of the hour.

That the camel is wholly unsuited to operations in-
volving daily movement (a fact well known to those
who have studied the peculiarities of the animal) has
been conclusively proved by the Afghan and Soudanese
wars. Slow feeders, they require frequent days of rest to
give them time to graze. While their capacity for
storing up water, and their indifference to heat, has
made them the ships of the desert, their delicacy of
constitution renders them unfit to withstand a great
strain on their energies or to undergo hardships. For
slow steady work at the base they are well adapted ;
with a rapidly moving column they are out of place.
In the Kuram valley, during the first portion of the
Afghan War, 9,496 disappeared out of a total of 13,480,
statistics that cannot but condemn them as transport
animals where rough work has to be accomplished.
The contrast between the horses and the camels during
the trying operations in the Bayuda Desert was very
marked. The horses, although reduced to prostration by
want of water and fatigue, stood the severe test and
soon recovered ; the camels succumbed[1].

A line destined to be of momentous strategic importance is the Trans-Siberian railway from Moscow by

[1] Callwell, *Campaigns since* 1885.

Samara, Ufa, Omsk, Tomsk, Irkutsk, Chita, the north of
the Amur River and the east of the Ussuri ; with proposed lines to link it with Port Arthur and Peking by
Tsitsihar, Kirin, and Mukden.   Yet it has its drawbacks,
and many are of opinion that these lines need give the
British little uneasiness, at any rate for a long period to
come, if they only adopt a resolute and fixed Eastern
policy, not to be varied by any changes in party government.   Vacillation is fatal to strategy, however favourable
may be the strategical conditions.   It is said that railways are the true civilisers of the world, but this depends
on what kind of civilization follows the advancing line.
The Soudan line has already helped civilization, inasmuch as the rule of the Khalifa was utterly ruining the
dwellers by the banks of the Upper Nile ; the Canadian
Pacific opens out new realms to enterprise and cultivation which were merely a wilderness, and none of the
original inhabitants have suffered in consequence.   But
planting so-called Russian civilization throughout all
Siberia will probably not elevate the folk who now
dwell between the Ural and the Lena, however much
it may promote the interests of new settlers.   Strategy
and tactics were too much for the Zulus and other
Kafirs, but they are far lower now in the scale of
human excellence than they were in 1833. So little
impression have the Portuguese rulers of East Africa
made on the natives under them that horrible atrocities
have quite recently been enacted under the very eyes
of their *capitaõ mor*. The people of Gazaland were in
such dread of the fate hanging over their next generation
if put under Portuguese influences that in 1891 two in-
dunas were sent to England to beg that Gungunhama's

territory might be placed under British protection, but in
vain. The Central Pacific from Chicago to San Francisco
certainly opened up new territories for the development
of the families and the gratification of the avarice of the
white man, but the red men were ruined. Corn and
domestic cattle superseded their buffaloes, which moreover were wantonly destroyed by monstrous battues.
The tribes of Indians whose ancestors roamed only two
generations ago in prosperity and freedom between the
Alleghanies and the Rocky Mountains are now practically doomed; their grasping neighbours will not even
keep their contracts, and have not scrupled to make
inroads on their "reservations." Indeed the reserve of
the Utes in Colorado has been invaded as lately as May
1899.

The British expedition to Abyssinia was merely an
affair of marching over mountains and subsisting. Half
of King Theodore's troops were disaffected, and in no
case did they make any stand worthy of description
against the British.

The battle of Magdala was won by Sir R. Napier's
troops with the utmost ease, but supplying the army on
its march from Annesley Bay, January 4th, till its re-
embarkation June 2nd, 1868, was the difficulty. In this
campaign, for the first time since the days of the
Romans, elephants were used by European troops in
Africa for purposes of military transport. The cattle
employed were 45 elephants, 7,417 camels, 12,920 mules
and ponies, 7,033 bullocks, and 827 donkeys.

It is said that during their march to Magdala the
British followed in the footsteps of Ptolemy Euergetes,
who started from the port of Adulis, close to Malkotto

on Annesley Bay, and having conquered most of Tigré returned to his place of disembarkation. The British force numbered 16,000 troops and 12,640 followers. The line of march was for 400 miles to Magdala over a mountainous and unknown country. The only previous instance of an invasion of Abyssinia by European troops was in 1541, when 400 Portuguese were sent to the assistance of the Emperor whose kingdom was overrun by the "Moors" (possibly to be identified with the Gallas of to-day). It is supposed by some that these troops entered Abyssinia from Zeila, Amphila Bay, or some point much to the south of Massowah, but it is much more probable that they entered from Massowah or Arkiko, four miles south.

Abyssinia is a factor which cannot be neglected by any Power interested in the future of Egypt, and in maritime supremacy in the Red Sea. Great Britain certainly was friendly to the development of Italian influence, but Menelek proved as terrible a foe to the Italians as had his predecessors to Mohammedans from the north and south in the middle ages, and to the Egyptians in this century. Under an enterprising leader a numerous, well-armed, and well-supplied Abyssinian force might well be as serious a danger to the dwellers in the Nile valley as was the Mahdi himself.

France and Russia have of late displayed considerable interest in this ancient state of Ethiopia, with which they had relations from the days of the Byzantine Empire, but more with the object of thwarting British influence than because of any immediate gain to themselves or benefit to the Abyssinians. It is admitted by Russian writers that they desire to become paramount in Abyssinia so as to threaten our communications with India through a base on the Red Sea littoral and near the Straits of Bab-el-Mandeb, where the French have already Sahalo and Obok and almost all the Gulf of Tajura[1]. Fedoroff advocates the following policy of grasping the hand which the orthodox Christians under the Negus have so often extended for help to Moscow. "Our engineers, mechanics, and craftsmen can there build up an industry, and our enterprise and skill can secure a hold on both the import and the export trade. This would give us command of the economic life of Abyssinia. The appointment by our Holy Synod of an Abyssinian Metropolitan would surrender the country to us in a religious sense. The despatch of military in-structors and the formation of a body of Abyssinian troops under the guidance of Russian officers would place in our hands the armed force." He goes on to prove that the results of this policy would be every-where apparent. It would doubtless have far-reaching consequences and would forthwith affect the strategic geography of our empire at large. The Abyssinians under these conditions would certainly give employ-ment accompanied with much risk to the Anglo-Egyptian army, but—worse still—Russians could easily acquire one of the southern ports. "This port would have for us a deep significance as a coaling station, and still more as a source of perpetual threat to the welfare and consequently the power of England. Our torpedo boats and cruisers could thence at any moment close to English merchant steamers the trade route through the Red Sea, and this would of course cut in half the trade

[1] See Col. Gowan's translation, *Journal of U.S. Institute*, p. 1271.

relations of England. The possibility thus afforded to us of closing this route would compel the proud Britons to considerably lower their tone, and it might aid us in the settlement of the Eastern Question in a sense that is to be desired by Russia."

Thus it is honestly admitted that the old realm of Prester John is to be used as a powerful weapon against British prestige and British commerce, both in Africa and Asia. Of course if the British gained the confidence of the Abyssinians this plan would fall through, for notwithstanding the comparative security and the complete success of Napier's expedition, the country is a splendid theatre for able defensive operations. To quote Colonel Gowan's words, "The whole country consists of elevated mountain plateaux (ambas) which are abundantly supplied with water, and cultivated fields, which yield two harvests in a year. Such natural fortresses (that of Gunib for example) are separated from each other by deep gorges and river beds, along which giddy precipices afford the sole means of communication. In many places access to the summits of these ambas is possible only with the aid of scaling ladders. It is true that there are fairly good pathways between the capital of the petty sovereignty of Gondar and Massowah, and from both these points to Magdala, but these wretched tracks are not practicable for either wheeled transport or for field artillery. The conduct then of offensive warfare in such a country is evidently not an easy undertaking."

Be this as it may, the fact that discussions on such points are of frequent recurrence in foreign political and military centres proves that ignorance of strategic

geography on the part of our statesmen may involve
our nation in irretrievable disasters.

It is a proud reflexion that with the solitary exception of South Africa, where the accident of Boer rule
was allowed to check and alter all our traditional policy,
the opening of new roads and new railways, and the
strategic advances of the British, have in every case
been followed with such a distinct and rapid improvement in the status of semi-civilized or semi-savage
peoples, that the principal danger is that the unwonted
peace, security, and wealth which ensue may lead to
over-population.

CHAPTER XI.

FOR the future the strategist in Europe will be tied to railways; the lines of military operations over which rails are not laid are diminishing yearly. But as these are identical in most cases with the old roads, and must of necessity go through the same passes and defiles, it may be well to consider the main roads which armies have followed.

The gates of Eastern Europe, excluding sea passages, are from Siberia by Ekaterinburg into the East of Russia, by the line now taken by the Trans-Siberian Railway to Samara and Moscow; by Orenburg; and *viâ* Derbend through Georgia, and by the Dariel Pass in the centre of the Caucasus into southern Russia. The Turks first landed in Europe at Gallipoli. Before conquering Constantinople they found their way to Adrianople and to the Danube by the well-known passes of Slivno or Selina, Shipka, Trojan, and Etropol; and after the fall of the old Eastern Empire they went by Servia, which had been ruined at the battle of Kossova in 1389, to Belgrade, "the Gate of the Holy War," and to Buda-Pesth. The road as far as Vienna was twice traversed by Moslem hosts, but on each occasion they failed before its walls.

Gourko passed the Balkans between Orkanie and Sophia in eleven days, during which he lost three generals, twenty-nine other officers, and 1,000 men. Before his movement began, during a great snow-storm, December 18th—23rd, he lost 2,000 men in one night. During the same storm the 24th Division in the Shipka Pass lost 6,000 men from the same cause. After the Danube bridges were carried away they had to depend for six weeks on ferry-boats or transit over unsafe ice. They practically lived on the country, but by this time Bulgaria was nearly exhausted. The campaign was made on hard bread and cattle, driven along on the hoof; no other rations, small or great, were available. The troops had had no change of clothes since the beginning of the campaign, and their ragged and patched condition was deplorable ; even their boots were patched with canvas. The Turkish dead were stripped for clothes, and Skobeleff on the road to Adri-anople compelled the peasants to bake soft bread, but these were mere makeshifts. Gourko's artillery was left north of the Balkans. Skobeleff and Mirza from Shipka had cannon, but their horses suffered severely. The cavalry, being well in advance, procured a fair amount of forage, but the fine artillery horses were mere skeletons when they reached San Stephano, the roads being covered with smooth ice, on which they slipped and suffered severe injuries. The cavalry horses had consumed the forage in the villages *en route*, and much of their desiccated food was eaten by the gunners. The baggage-waggons were left north of the Balkans.

Officers' baggage, very much curtailed, was carried on pack-horses; the men carried their baggage and 20—40 rounds of ammunition in their pockets, and their rations on their backs, the knapsacks having been all left behind. The men frequently had to fight without having tasted food for twenty-four hours. But the peasants of Russia are used to hardships, and no complaints were heard. Their adversaries rivalled them in self-restraint and powers of endurance, but the seeds of typhus and typhoid were laid in the Balkans, and when a milder clime was reached, disease raged amongst the long-suffering troops within sight of Stamboul. The sappers misled Gourko as to the difficulties of the Orkhanie road, they said it was practicable for artillery, but the best parts had a slope of one in six, and the worst a slope of one in three. The horses were useless. The guns were hauled along, or let slide down ice-clad inclines, by drag-ropes. The celebrated Preobraghensky 1st Regt. of the 1st Division of the Guard maintained its old prestige by leading in this appalling enterprise.

Invaders of Hungary enter by the Rothenthurm Pass, south of Hermannstadt, which was the route of the Russian General Lúders in 1849; by the Tömöser Pass south of Kronstadt; by Vereczke, the gate of the Magyars; and by the valley of the Bogrod and that of the Vistula.

The *trouée* of the Oder leads into Moravia, and in the Seven Years' War 1756–1763, in 1813, and also in the Seven Weeks' War of 1866 the passes used in the north of Bohemia were Nachod, Trautenau, Reichenberg, Pirna, and Töplitz.

Armies invading Italy would move from Vienna by

the Semmering Pass, thence to Villach, and by the Tarvis Pass between the Carnic and the Julian Alps to Udine. Napoleon went by Tarvis as far as Leoben in 1797. From Bruck on this road there is a line by Laibach to Trieste and Fiume; while Innsbruck is joined to Switzerland through Arlberg and to Vienna by Kufstein.

A road was made from Innsbruck through the Brenner Pass by the Romans, and it was for some centuries an avenue for barbaric inroads, and afterwards the main route of commerce between Venice and central Europe. It was of great significance in Bonaparte's campaign of 1797. The Stelvio Pass was opened up by the Austrians for strategic purposes; the Splügen is celebrated for its passage by MacDonald in 1800. St Gothard, connecting the valley of the Reuss with the valley of the Ticino, is a noted pass, whence the old road from Italy goes to Altdorf, Zurich, and Schaffhausen, where it joins the principal lines which turn the Black Forest. Suwarrow used the Gothard in 1799. Moncey joined the First Consul at Milan from Moreau's army at Schaffhausen by this road in 1800.

The Simplon is now a fine road, and other important passes are the Grimsel, Gemmi, Bernardino, Septimer, and Bernina. Altogether there are some forty commercial highways over the Alps.

A case in which a railway has not superseded a road for military purposes is the St Gothard line. Even supposing the tunnel not to be obstructed, its services in transporting troops would be small. It can only be used by demi-trains, but 100 complete trains are required for a *corps d'armée*, therefore in the tunnel 200

trains would be required, and when the special engines
are considered, the reduced speed by reason of slopes
and curves, and the absence of disembarkation stations
and sidings, only ten trains daily could be sent on.
Therefore it would take twenty days to transfer a corps
of 30,000 men with its impedimenta, from Bellinzona to
Lucerne or *vice versâ*. In good weather therefore the
old. road would probably be the quicker route.

The Pusterthal defile connects the Tyrol and Vienna;
the Maloja the valley of the Inn with Italy.

The Simplon, with a road made by Napoleon, gives
passage from the valley of the Rhône to the valley of the
Ticino. The Furka Pass connects the valley of the
Reuss with that of the Rhône. The defile of St Maurice
is a formidable strategic position, the proper fortification
of which would render Switzerland impregnable from the
direction of the Rhône valley.

The Great St Bernard leads from Martigny to Aosta
and Turin, and it was by this pass that Bonaparte turned
the right flank of the Austrians in 1800.

The Little St Bernard, Mont Cenis, Genèvre, Tenda,
and La Corniche passes lead from France into Italy, and
have been traversed by armies from the days of Charles
VIII—the first French monarch who interfered to any
considerable extent with the affairs of Italy—to those
of Napoleon III. The Radstädter Tauern connects the
Drave with the Salza. The Rottenmanner Tauern is
between the Ems and the Drave.

North of the Alps the defiles are along the line from
Geneva to Lyons, across the Jura at Pontarlier—scene
of the surrender of Bourbaki's army in 1871—and
Porrentruy. The gap of Burgundy from Basle into

France near Belfort was used by the Allies in 1814; it is between the Jura and the Vosges. Across the latter mountains there are numerous roads from Mulhausen, Colmar, Schlettstadt, and Strasburg. That from Strasburg to Épinal by the pass of Schirmeck was used by Von Werder when, after the capture of Strasburg, he proceeded to try his fortune in the directions of Dijon and Besançon. The gap of Saverne is on the main line from Strasburg to Paris. Bitsch commands another route through the northern Vosges. All important Rhenish towns from Strasburg to Arnheim are centres of old roads and railroads leading into the valleys of the Moselle, Meuse, and Seine, and eastward to Leipsic and Berlin, Warsaw and Königsberg, and St Petersburg, Wilna, Moscow, and Kiev.

The direct line between Paris and Berlin passes by Maubeuge, Namur, Liège and Cologne. But neither a French nor a German army would take this line, inasmuch as they are now face to face, and almost touching each other, on the frontier of Lorraine. Moreover, physical obstacles would render it impossible for an army in Belgium and one in Lorraine to keep touch.

In the Roman days Germanicus passed the lower Rhine because central Germany was covered with forests and the cantonments of the barbarians were in the north of Germany.

For the same reason Charlemagne advanced against the Saxon Confederacy by the lower Rhine: the army of the Sambre and Meuse crossed and recrossed at Düsseldorf in 1796. The northern army of the Allies also advanced by Düsseldorf in 1814.

There are numerous roads through the northern part

of the Black Forest. In the southern part the principal defile is Höllenthal, near Freiburg. The *trouée* south of the Danube is between the Wutach River and the Lake of Constance, and by this route the French marched into Germany in 1796 and 1800. There are no difficult passages in the gates of Westphalia or the gates of Thuringia. The "Rennstiege" of the Thuringian Forest deserves mention—it is the oldest military road in Mid-Europe, and (according to Freytag) was in existence before our era. It extends from Eisenach to the Fichtel Gebirge, and although now in desuetude and decay, it was undoubtedly once a great road. It leads right along the watershed of the forest.

The passes from France into the Iberian Peninsula that can be used by armies are few. On the west the road by Bayonne, St Sebastian, Burgos, Valladolid, and thence to Madrid or Lisbon, was used by Napoleon and Wellington. On the east the route from Perpignan to Barcelona, Valencia, and Carthagena has been followed by many an army, from Hannibal's to Suchet's. There are also minor roads, Lerida by Séo d'Urgel, Zaragoza by Jaca and Urdos into France, and lastly the pass of Roncesvalles, used by Charlemagne.

There were two main roads from Portugal into Spain during the Peninsular War. The northern went from Lisbon by Almeida through the fortress of Ciudad Rodrigo to Salamanca, thence bifurcating to Madrid and to Burgos, *viâ* Valladolid; while the southern road passed from Lisbon over the Tagus and through the great fortress of Badajoz, whence it stretched south-east into Andalusia and north-east to Madrid. Hence the strategic importance of the two fortresses, both of which

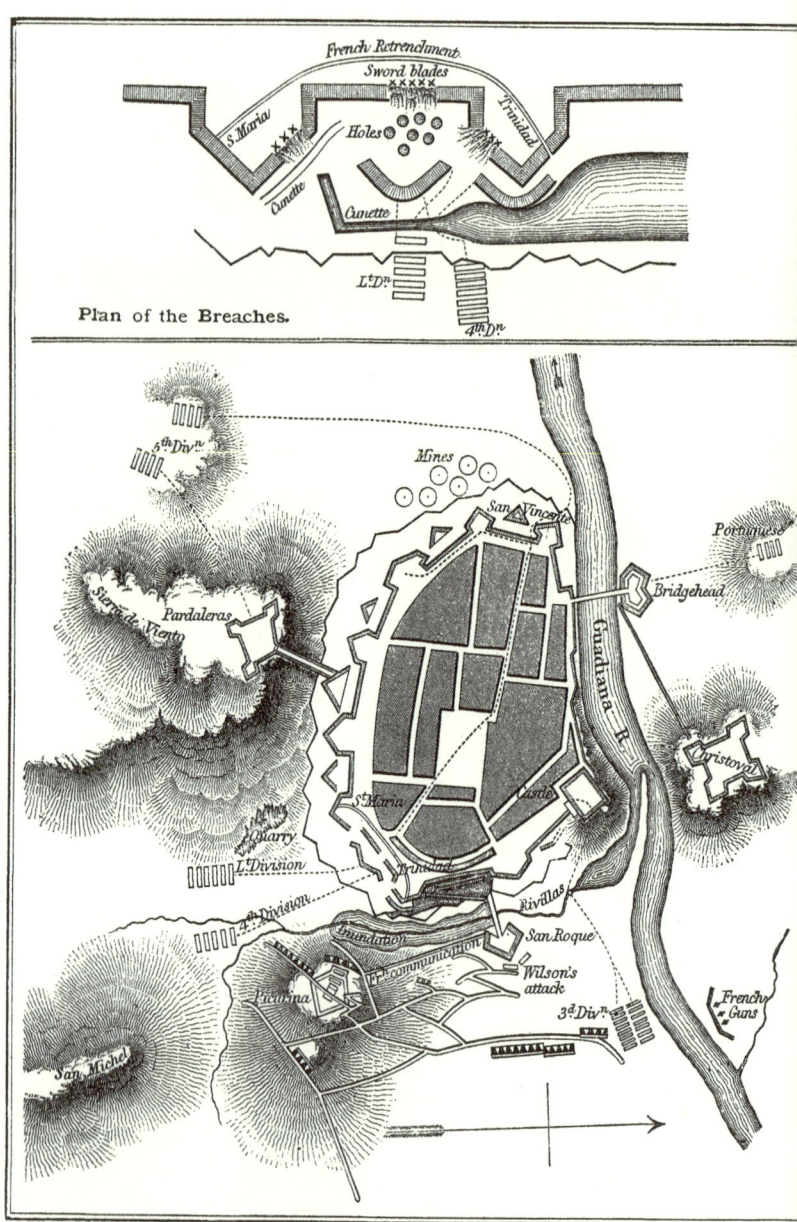

French Retrenchment.

Sword blades

S. Maria

Trinidad

Holes

Cunette

Cunette

Lt. Dn.

4th Dn.

Plan of the Breaches.

5th Divn.

Mines

San Vincente

Portuguese

Sierra de Vienta

Pardaleras

Bridgehead

Guadiana R.

Cristoval

St. Maria

Castle

Quarry

Lt. Division

Trinidad

4th Division

Rivillas

Inundation

Communication

San Roque

Picurina

Wilson's attack

San Michel

3d. Divn.

French Guns

STORMING OF BADAJOZ IN 1812.

Wellington was compelled in 1812 to take by storm at all costs. The scenes at the assault of Badajoz were the most terrible in the annals of the war. At the main breaches 2,000 British troops fell within the space of a hundred square yards in one hour.

The road into Greece used by the Turks in the war of 1897 was by Salonika, Elassona, Larissa, Velestino, Pharsala, Domoko, and Lamia—near which is Thermo-pylæ; on this line they had a secondary base at Monastir in Albania. There were also operations on the line Janina to Arta, and this road would have led to Lepanto. The Turks, next to the British, are the most skilful adepts in moving troops by sea in Europe, and could have crowded troops into Salonika if need had arisen, but the resistance of the Greeks collapsed without any necessity for great efforts of this kind.

During the Roman wars the mountains of the Olympian chain were the scene of singular and interesting tactics. Antiochus III from Syria seized Thessaly, and as the Consul Cecilius advanced against him, he took post at Thermopylæ, strengthening the entrance of the defile by redoubts and a double ditch. Slingers and archers occupied these works, and in front the phalanx formed a rampart bristling with pikes, his right towards the sea being covered with elephants. But the elder Cato, who served on the staff, turned his left, and descending on the rear of the Syrians caused them to retreat with precipitation; the elephants, however, covered the rear in a narrow defile and completely stopped the pursuit.

In the war against Perseus, 171—168 B.C., the Romans were again in great difficulties in this Olympian chain.

The Consul Marcius Philippus, wishing to avoid any conflict with the Macedonian troops who watched the defiles, marched through almost impracticable by-paths which the enemy thought it would be useless to guard. His army in these passes, amidst sharply escarped hills, suffered severe privations, but what caused most delay was the difficulty in persuading the elephants to descend by the steep slopes. They were frightened by the precipices and, weary of a succession of difficult obstacles, tossed off their drivers, and gave vent to their rage in screeches which frightened the horses. They caused the army as much trouble in every defile as a surprise by the enemy till the consul devised a plan of luring them on to moveable bridges, and thus transferring them across the deepest gullies.

It is always dangerous to risk an important enterprise in any defile, the force may be taken by a surprise, and then escape is only possible if some other defile be at hand. Maurice of Saxony and his mercenaries in 1558 marched from North Germany into the Tyrol, taking by surprise the fortresses on the road to Innsbruck, driving the Emperor Charles V to a precipitous flight over the Carnic Alps, and enabling him to dictate the terms of the Treaty of Passau. In this case the number of the assailing body was contemptible, and a little presence of mind would have taken its measure and baffled its purpose. But the danger of surprise consists in the disturbance it causes in the mental vision of the men surprised. Coolness and the power of reflection are replaced by panic and dismay. This example of a surprise inculcates the necessity of attention to every detail, even the smallest.

The Romans practically surrounded their Empire, whether by the banks of the Tigris, the Danube, and the Rhine, or north on the line of the Solway and Firth of Forth, with a series of military positions along which the legions placed posts, their head-quarters being frequently behind in some central camp or colony, whence they could with ease reinforce any part of the threatened line. Rapid communication with every part of the Empire was secured by the admirable military roads, which in every case accompanied the progress of their arms, linked the farthest settlements with the centre at Rome, and were at once magnificent avenues for commerce and strategic lines of operations. No question of cost was allowed to interfere with their excellent structure and perfect preservation. The densest forests were cut through and the most rapid torrents crossed, witness the sublime bridge of Trajan at the Iron Gate of the Danube near Orsova. For centuries the Empire expanded and the frontier was constantly advancing, but there came a time when the course of Empire stopped, and forts, walls, and vast dykes, which tradition later ascribed to diabolical ingenuity, took the place of powerful armies ready for the fray.

The end of the Empire was then at hand. The merely defensive attitude behind lines of forts is a sure sign of national decay; the descendants of the Cæsars sank into timidity when victory seemed uncertain, and before very long the roads became easy avenues for barbaric invasions. Yet they are still topographical and historic landmarks of practical importance; indeed they were built for immortality, and their milestones—starting from the Golden Milestone in the

Forum on which all distances were marked—are still frequently found in Europe, Asia, and Africa.   No wonder
that Rome appeared to the natives beneath her sway
eternal as well as all-powerful, so that the fall of the Imperial city seemed to mark for them the end of all things.

All writers on European strategy find their labours
lightened by a good classical atlas, and the solution to
many difficult problems of modern history will be found
in the geographical records of the civilizations before
the Christian era, of which, in his *History of Ancient
Geography*, Mr Tozer has given an admirable summary.
The railway route of to-day from Constantinople to
Central Europe is by Adrianople, Philippopolis, Sophia,
Nish, to Belgrade, and thence to Budapest and Vienna,
and the traveller who books his seat in the " Orient
Express" from the latter city to the Turkish capital
follows throughout his journey the track of a Roman
road ; while if he passes on into Italy he will soon be
in the Via Aemilia itself by Verona, or by Aquileia.
Attila took the latter road with disastrous consequences
to the inhabitants.

The British or Saxon traveller who—either as a
victim for a Roman triumph or a pilgrim of the faith—
made his toilsome way to the centre of victory and
religion, having crossed the Channel from Richborough
to Gesoriacum or Boulogne, proceeded thence by Amiens
(Samarobriva) to Châlons sur Saône (Cabillonum), thence
to Lugdunum or Lyons—which as the confluence of the
Saône and the Rhône was in the days of Agrippa a
leading strategic centre even as it is now—and by Milan
and Placentia to the Æmilian and Flaminian Ways.
From Lyons a great road branched off to Arles and the

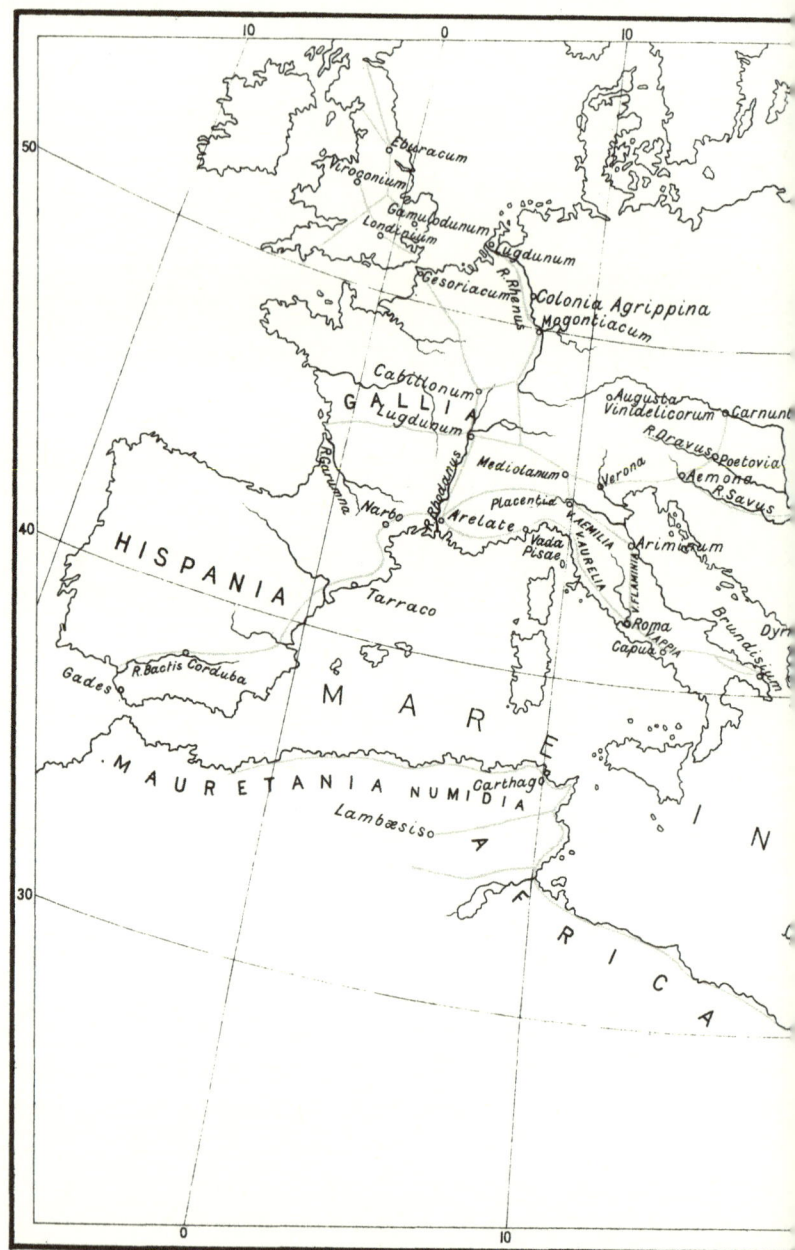

THE CHIEF LINES OF ROAD IN THE ROMAN EMPIRE

30 40 50

50

DACIA

dunum

R. Ister

ESIA

Serdica

ATIA

ia

Thessalonica

PONTUS EUXINUS

Byzantium

Ancyra

ASIA MINOR

40

30

Tarsus

Antioch

Tigris

MESOPOTAMIA

R. Euphrates

Seleacia

30

E R N U M

SYRIA

Palmyra

Alexandria

ÆG

Y

R. Nilus

P

T

U

S

Myos Hormos

Coptos

Berenice

40

30

Edwin Wilson, Cambridge

Mediterranean, another went by Limoges to the mouth of the Garonne, and another to Rheims, which has been equally important in the Hundred Years' War, in the Napoleonic wars, and in the last Franco-German wars. Another road was from Lyons to Vienne, and by the Little St Bernard to Aosta, Ivrea, and Mediolanum (Milan), whence a continuation went by Turin and the Durance to Arles.

The Via Flaminia skirted Mt. Soracte, passed the Apennines, and went to Rimini (Ariminum): under the name of Via Aemilia it was continued to Piacenza—the celebrated passage of the Po by which Napoleon turned the Austrians in 1796 and 1800—and Milan, afterward the capital of the Lombard Kings and Napoleon's capital when he ruled the kingdom of Italy. By the Great St Bernard a road went through Helvetia to the Rhine near Basle. The four roads through the Alps known to Polybius were by the Riviera, Mont Genèvre, the Little St Bernard, and the Brenner: the last-mentioned defile went to Augusta Vindelicorum or Augsburg, which maintained its commercial position through the Middle Ages, and its strategic importance till the wars of Marlborough and Moreau. From Verona roads stretched by Laibach on the one side to Carnuntum on the Danube, and on the other to Singidunum (Belgrade).

Carnuntum was one of the chief Roman fortresses on the Danube, and a station of the Danubian fleet. When Marcus Aurelius was fighting the Marcomanni and the Quadi he made it his head-quarters. It was finally destroyed by the Hungarians. Its site may now be traced by ruins near Deutsch Altenburg, east of Vienna (Vindebona).

The Via Aurelia went by Pisa, and was continued under the name of Via Aemilia Scauri to Ventimiglia, Monaco, and Cimiez, where it branched off to Fréjus and thence to Narbonne, and crossed the Pyrenees to Gerona —the scene of fierce warfare against Augereau, Mac-Donald and Suchet in 1809-1812,—Tarragona, Valencia, and the Jucar, the theatre of the British operations under Murray and Bentinck in 1813. It branched off to Corduba and Hispalis (Seville), and terminated at Cadiz (Gades).

The quickest route to Byzantium was along the Appian Way to Capua—where Hannibal was reported to have let his opportunities slip by in inglorious luxury and repose—and Beneventum, and by Tarentum or Canu-sium to Brundusium, where the Adriatic was crossed to Dyrrachium and Apollonia, whence the Egnatian Way brought the traveller eastwards by Monastir and Pella, the birthplace of Alexander, to Thessalonica (Salonika) and by Philippi, where the Triumvirs defeated Brutus and Cassius, to Constantinople. The road from the Hellespont to the Tigris at Seleucia (so called from one of Alexander's generals) went from Abydos to Ancyra, the scene of Timurlane's great victory over the Ottoman Turks, and by the Cilician Gates to Tarsus. It then passed the Pylæ Amenides to Antioch, one of the last fastnesses of the Crusaders in Asia, and crossed the Euphrates by the bridge of boats at Zeugma. There was also another road from Antioch to Seleucia through Palmyra, which city was for centuries a leading centre of trade and art till overwhelmed, in spite of the energy of Zenobia, by the savage vengeance of Aurelian, 272 A.D.

A road ran along the coast from Antioch to Egypt, by which both Alexander and Bonaparte proceeded up the Nile to Coptos near Thebes, and thence to Myos Hormos on the Red Sea. A strategical road also linked together all the stations and seats of commerce on the North African shores, and in N.E. Gallia a road connected the various posts on the left bank of the Rhine from Mayence to Colonia Agrippina (Cologne) and to the mouth of the river near Lm (Lugdunueyden). This latter work anticipated the modern railways that connect the German fortresses, and was constructed for similar strategic reasons. The terrors of the Hyrcanian Forest, which stretched from the Teutoburger Wald where Varus lost his legions to Poland, stopped Roman enterprise in Northern Germany.

The operations of the French in the Alps in 1799 are frequently used as illustrations of the use of mountains for defensive purposes and of their defiles for offensive strategy, and especially turning movements. Suwarrow's enterprises in 1799 consisted of three distinct movements, in each of which he was foiled by reason of the ability of the French and the inferior energy of his allies the Austrians. Advancing into the St Gothard defile from Airolo himself, he sent detachments round by his right to Crispalt and Disentis. The French generals Gudin and Lecourbe retired, the former, moving by the Furka Pass and the Rhône glacier, the latter breaking down the Devil's Bridge. A shocking scene took place at this dreadful gorge, but at last the Russians crossed and drove the French back to Altdorf; but when Suwarrow reached the Lake of Lucerne he was stopped by the want of boats, and was therefore obliged to retire by his right in a most perilous journey by the Shächen-Thal towards Glarus.

No more heroic march than this has ever been undertaken. The Russians abandoned their artillery and baggage and advanced in single file, dragging their beasts of burden after them. The leading files had reached Mutten before the last had left Altdorf, and the precipices beneath the path were covered with horses, equipages, arms, and soldiers unable to continue the laborious ascent, but the Austrians again retired, and at Mutten the Russians were in the midst of French enemies. They attacked Molitor and drove him to the banks of the Linth, but he stood at Näfels and blocked their way in spite of the resolute attacks of Prince Bagration. At the same time the Russian rear guard was assailed by Massena, who was repulsed. After resting for a few days at Näfels, the Russians resumed their retreat over the mountains by Engi and Matt into the Grisons. Snow was falling, the tracks were obliterated, few stores were left, there was no shelter. The ascent by the Panix Pass was dreary and dangerous, the descent to Ilanz was even more risky by reason of the slippery frozen snow. After sleeping in snow without fire or covering they reached the valley of the Rhine with head-quarters at Ilanz. At Coire there was a liberal distribution of rations, and at Feldkirch the Russians reopened communications with the Austrians. Five thousand Russians perished during these marches in one fortnight.

The leading difference between the operations of Napoleon and MacDonald in the Alps as compared with the only failure in the glorious career of Suwarrow was that the combinations of the former were sound, and if the defiles could be passed their adversaries would be at a strategic disadvantage of a serious kind ; moreover, the facts about the routes throughout were known to both the French leaders, while the Russian general was quite misinformed about the topography of the district. Moreover, the Austrians and Russians in the Alpine districts in 1799 were on exterior lines, while Moncey and Napoleon in 1800 were on interior lines.

In Suwarrow's case sixty-six thousand allies, divided into several large sections, spread out in a wide semicircle from the Lake of Zurich to St Gothard and to Disentis, and attacked a superior enemy in a central position. Moreover, there were two fatal errors in detail. The road selected became a mere bridle-path at Taverna, south of Bellinzona, and came to an abrupt end at Altdorf, though Suwarrow's object was Schwyz. Moreover, the Austrians promised to have fifteen hundred mules at Taverna. After a delay of a few days only six hundred and fifty arrived, and it then became necessary to requisition the horses of the Cossacks for baggage animals.

Napoleon's strategy in May and June, 1800, was brilliant both in conception and execution, and his elaborate plans resulted in such complete success as to dazzle mankind, and to win the warm praise of General Bulow, who compares his foresight and quickness with the narrow views and hesitating manœuvring of the Austrians. But his passage of the Great St Bernard in May by St Pierre, Etroubles, and Aosta, though excellent from every point of view, was far from being as risky an operation as many other similar movements, and the passage, which lasted only three days, was not hampered either by the inclemency of an Alpine winter or by any military resistance ; indeed it was facilitated by the hospitality of the monks of the Hospice. But the experiences of Suwarrow previously in the St Gothard, and Glarus and Grisons was of the most appalling character. He had to fight both nature and man ; as had also MacDonald in his desperate movements in the Splügen and among the glaciers of Mt. Tonale, and in his long circuit by Pisogne and the valley of Sabbia and the Val Trompia into the Sarca valley. MacDonald concentrated between Coire and Tusis at the entrance of the Via Mala in the middle of December, and followed the road over the Rhine till he reached the village of Splügen, where the serious difficulties began. Avalanches swept his men over precipices from their track over the snow, which was only formed by the pressure of an advanced party of the strongest men and heaviest cattle of his army. On the night after the Hospice was reached—the 5th of December—all traces of the road were lost in snow, yet the summit was attained. Here an avalanche closed the descent, but MacDonald led the advance himself, sounding the unstable mass of snow with a long stick. The soldiers cut through walls of ice and snow, though suffering heavy losses by frequent avalanches before they reached Isola. Thence he was ordered to cross the Col Aprica by steep ascents and difficult roads into the valley of the Oglio. He then had to lead his troops in single file by a path through the snow to the summit of the Tonale. Advancing along a plateau 300 yards wide between two impassable glaciers, he found himself before

a triple line of Austrian entrenchments faced with solid
ice. These his men could not carry in spite of desperate
efforts, but as is usual in these cases, a turning movement by another path succeeded, and his left wing
entered the Engadine and seized Glurns.

He reorganised his force in the Val Camonica;
again he could not force the mountains in his front, so
he turned their southern extremity by Isogno. Thus
the Austrian line was turned by the upper extremities
of the Tyrolese valleys. As far as the obstacles of
nature are concerned this was the most marvellous
operation in European warfare in modern times. It is
but fair to say that MacDonald only ventured upon
such an enterprise in consequence of the peremptory
orders of Napoleon, who little thought that the operations of his lieutenant would gain more glory than his
own.

One of the most singular military episodes in Napoleon's career was the surprise and capture of Vandamme
at Kulm in the Pirna-Töplitz defile. After the glorious
victory of Dresden, Vandamme pressed upon the retreating Allies in the direction of Töplitz; in point of
fact he put himself astride the road, and had he been
victorious in the battle which followed, the Allies could
not have debouched from the Erzgebirge. Moreover,
as they were pursued by the Emperor and Marmont
and Murat, and had several thousands of waggons in
the narrow defiles, they would have been ruined. But
Vandamme became isolated, and was not supported by
any reinforcements, which he had been led to expect,
nor did Napoleon pursue with vigour the retreating foe.
Thus Vandamme's 23,000 men were attacked in front

by 60,000 Russians and Austrians, and being obliged to fall back, found Kliest's Prussian corps on the main road. After a terrible hand-to-hand conflict in the narrow defile 20,000 French were killed or made prisoners, amongst the latter being Vandamme. The rest escaped without arms over the mountains to Peterswalde.

There is always a danger of fighting with a defile behind the line of battle, whether it be a road through a mountain or a bridge over a river. If the battle be lost and the retreat be not orderly, wild confusion is the result, the troops will be hemmed in by precipices or forced into the river. Thus at the battle of Leipsic there was only one bridge over the Elster, but it was broken down before all the troops were across; the result was appalling loss, including Marshal Poniatowski, who was drowned. Again, waggons may block the road, as at Vitoria.

But if the line of battle be not too near the defiles, and if the troops, covered by artillery and cavalry, be carefully drawn away betimes, defiles become an advantage to a retreating army, as was seen at the battle of Sadowa, and often on the Mincio. Jomini says that the forest of Soignies behind Wellington would have been of much advantage had he lost Waterloo, as the British would have fallen into disorder, and the defiles would have retarded Napoleon's pursuit.

The rule is that no attempt should be made to carry a mountain defile or a strong position on a mountain crest by a direct or front attack if it can possibly be turned by a flank attack. This rule has been adopted from the days of Xerxes to those of Moltke, who did

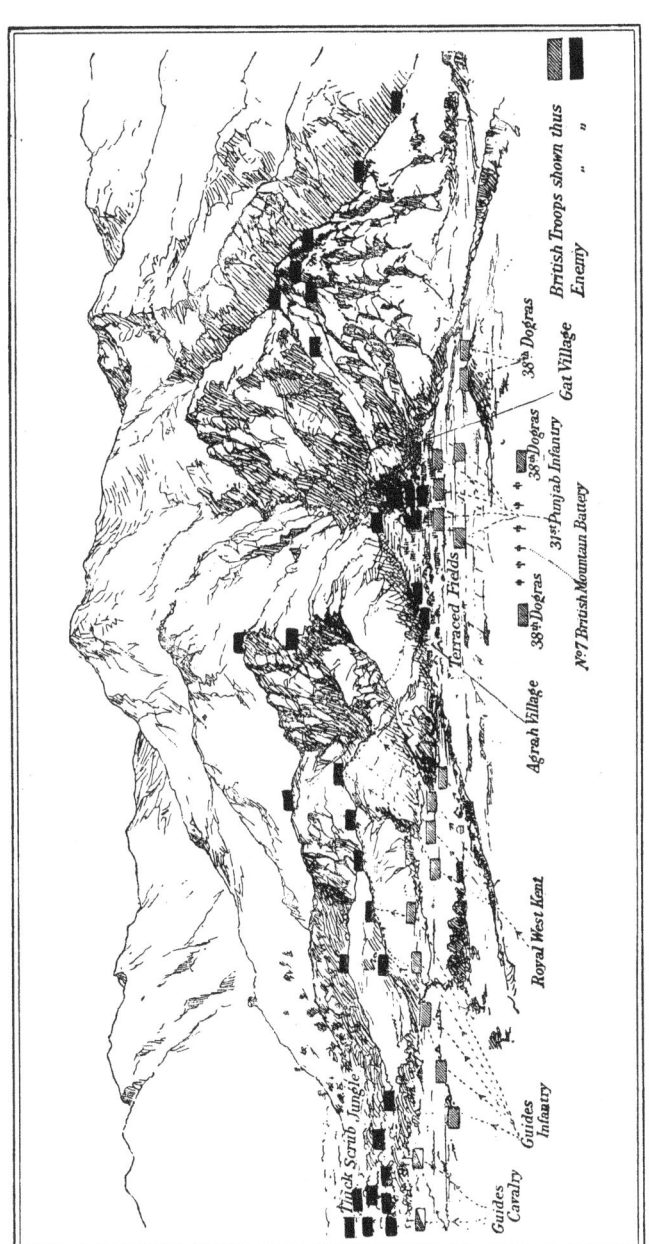

Guides
Cavalry

Guides
Infantry

Thick Scrub Jungle

Royal West Kent

Agrah Village

Terraced Fields

No 7 British Mountain Battery

38th Dogras

31st Panjab Infantry

38th Dogras

38th Dogras

Gat Village

British Troops shown thus

Enemy   "   "

SKETCH OF BRITISH ASSAULT ON AGRAH VILLAGE.

not attempt to carry Bitsch or Pfalzburg, but simply observed and marched round them. Suleiman's repeated attacks on the Russians at the Shipka Pass were among the worst blunders in Turkish history. The Russians used other minor defiles to turn the same pass after the fall of Plevna. The British, however, have had many fierce fights in defiles in recent years, as illustrated by the annexed plan of the assault on Agrah village.

Diebitsch, in the Balkan campaign of 1829, leaving a force to watch the Grand Vizier at Shumla, suddenly and secretly crossed the range with 30,000 men, and got supplies from the Czar's fleet under Admiral Greig in Burgas Bay. But for these supplies, and the skill with which he led the enemy to over-estimate his numbers, he would have entered Adrianople as a prisoner instead of as a victor.

It has been shown that in the Alps the Austrians repulsed the French front attacks on their positions, and that the French commanding the defiles repulsed the Russian attacks under similar conditions.

The actions of the American civil wars support European lessons; Sherman could not force one of Johnson's positions in the defiles between the Tennessee and the Chattahoochee in 1864, nor do the recent actions at Dargai by Generals Lockhart and Yeatman-Biggs lend credit to any different theory. Much controversy exists as to whether in this case a turning movement was possible. During the first battle the turning force under Brigadier-General Kempster did not arrive till late, and the road was so bad that no mules could accompany the men. The position was nevertheless carried, but afterwards abandoned. In the

second battle for the same position, though the enemy
was led to expect a flank attack, General Yeatman-Biggs
ordered a front attack. The result was a fierce engagement and the temporary check of some regiments. But
between Dorsets and Derbys, Sikhs, Gurkhas, and
Gordons the fight was well won at last. Much difference of opinion exists as to this battle. The attacking
general manifestly believes that he had no option, as
the movement to turn the enemy's left rear was an
impossibility ; on the other hand Sir William Lockhart
holds that it was possible and would have been decisive.
But traversing the Chagru defile under fire would certainly have been a serious enterprise, almost as bad
as MacDonald's over Mt. Tonale ; it is a narrow gorge,
sometimes little wider than the river bed, winding
between cliffs often rising perpendicularly 300 or 400
feet on each side. On the other hand semi-civilised
people or savages do not possess artillery, which is so
helpful to Europeans who command defiles, and further
a flank movement produces much more effect on Afridis
than on a corps like Vandamme's. No savage nation has
a good system of reserves, and against Asiatics the line of
the flank attack may be separated from the front attack
to a much greater extent than in Europe ; the enemy
has no reserve in hand to meet the turning movement,
nor can half-drilled troops change front left or right
with rapidity. Again, European leaders are skilled in
the use of detaining-forces, and all of them know well
how to hold back one attack with a small force while
concentrating against the other, and can execute quickly
counter-attacks, as at Austerlitz and Salamanca.

CHAPTER XII.

HISTORIC LINES OF INVASION.

THE name of Alexander, Philip of Macedon's "god-like son," has cast a spell over the imaginations of mankind, and the more carefully we study the map of the Mediterranean coasts, the seats of great monarchies mentioned in Holy Writ, and the confines of our Empire in India, the greater our admiration for his genius in war, his valour and his patience, as well as for his love of science and letters.

In the Romances of the Middle Ages he was a familiar figure, no old historic story was complete without some reference to his exploits; he figures in Spenser's *Faery Queen*, our Puritan poet Milton invoked his example so that "captains and colonels might spare the house of the author of *Comus* even as Alexander amidst the ruins of Thebes respected the house of Pindar," and the royalist Dryden in his immortal lyric makes him, under the fascination of St Cecilia, display all the rapidly changing transitions of his exalted and fascinating if capricious and unscrupulous personality. Albania attracted Byron because it was the home of Iskander; to Napoleon he was the lode-star; and the routes from the Eastern Mediterranean to the Punjab, though many had been opened up before his time, are mainly of historic interest because of his operations. The chivalrous and dignified chiefs of Kafiristan are proud to believe that his blood flows in their veins. No small number of the strategic points from the Helles-pont to Lahore are simply landmarks set up by his intuitive skill in mastering the relation of geography to war. But the generalissimo of Europe against Asia was not only a soldier; like all his most illustrious successors, from Cæsar to Trajan, and from Timurlane to Napoleon, he was also a student and a philosopher. The favourite pupil of Aristotle recognised that the road to knowledge was the road to victory; he knew that, as Spenser says—

> "Abroad in arms, at home in studious kind,
>   Who seeks by constant toil shall soonest honour find."

Every aspect of the effect of sea power was exem-plified in the methods whereby he linked his army with Greece, and gained new bases in the Levant before returning into the basin of the Euphrates and thence into Central Asia. His operations in the old centres of Assyrian and Persian power were eagerly followed by Moltke when engaged in the Turkish Service. From Rumkela, a town cut out of the rock, formerly the old Roman castle of Zeugma, the illustrious German, who was destined more than thirty years later to lead across the Rhine hosts more numerous than ever started from Babylon or Persepolis, wrote to his friends, " The Euphrates here reaches its extreme western points, and in former times it was crossed by a bridge, which was perhaps the reason why the Romans founded a colony in so impassable a neighbourhood. One starlight night

I stood amidst the ruins of Zeugma. Below in the rocky defile the Euphrates glistened, only the noise of its ripple breaking the stillness of the night. Before me in the moonlight passed the wraiths of Cyrus, Alexander, Xenophon, and Julian, who all, from this selfsame stony spot, once looked down across the stream upon the land of the Chosroes—a spot which time and nature cannot change." At this epoch (1838) the Egyptians were trying to conquer Asia Minor. Moltke reached the narrow defile through the Taurus which links Asia Minor to Syria, and at once recognised the strategic importance of Alexander's march.

From the days of Alexander to those of Ibrahim Pasha (2170 years) these mountains have played an important part in the march of armies, and even more so in the commercial intercourse of nations. These passes through which European armies had often advanced against Persia, India, and Egypt, had now to be closed against the latter nation, which threatened Europe with invasion just as it had five centuries before.

Alexander was first distinguished when eighteen years old in the service of his father at Chæronæa 338 B.C. On his father's murder in 336 he was elected as general of the Greeks against Persia, but he was first obliged to secure his communications by a campaign against the Thracians of the Balkans and against some of the Gatae, in pursuit of whom he crossed the Danube in 335. He returned to Greece, and having destroyed Thebes, which had rebelled in his absence, marched to the Hellespont with only 35,000 soldiers, confident that no number of Asiatics could resist his well-trained and highly disciplined army. At the river Granicus in 334

he defeated the Persians; his naval policy after this
victory has already been described. Next year he
collected his forces in Gordium in Phrygia, and, cutting
the mystic knot, set out for Issus on the boundary of
Syria. Here he inflicted a disastrous defeat on Darius.
In point of fact, by not occupying in strength the Pylæ
Amenides from Cilicia, the Persians almost courted
defeat. He then displayed the nobler traits of his
character in his generous treatment of his adversary's
family. He turned into Phœnicia, and after the fall of
Tyre received the homage of Egypt, founded Alexandria, and in the Lybian desert was welcomed as a god
by the priests of Ammon.

In 331 he set out to fight against the new levies of
Darius, crossed the Euphrates and the Tigris, and came
up with the grand Persian army at Arbela in the plains
of Gaugamela. This battle is used by Lord Bacon to
give point to his theory "that a man may truly make
a judgment that the principal point of greatness in any
state is to have a race of military men."

"The army of the Persians was such a vast sea of
people as did somewhat astonish the commanders of
Alexander's army, who came to him therefore, and
wished him to set upon them by night; but he answered,
'He would not pilfer the victory,' and the defeat was
easy[1]."

Babylon, Susa, and Persepolis surrendered forthwith.
In 330 he traversed Media and crossed the Paropamisus
(Hindu Kush) Mts. into Bactria, where he executed
Bessus, the murderer of Darius. He further dared to
cross the Oxus and Jaxartes and punish the wild Scythian

[1] Bacon's *Essays*, XXIX.

tribes to the east of the latter river. His policy was very different from that of the British ; they prefer isolation, he encouraged his soldiers to marry Asiatics, and set the example himself. He crossed the Swat valley and the Indus near Attock in 321, after an admirable display of tactical ability in mountain warfare similar to that which the British have so frequently undertaken since 1838 in the same part of the world. He traversed the Punjab, and defeated Porus on the Hydaspes (Jhelum) and marched as far as the Hyphasis or Sutlej, having treated Porus with magnanimity.

His troops would go no further ; he therefore took ship down the Indus with part of his army, while two divisions marched along the banks. Arriving at the Indian Ocean, he sent Nearchus with the fleet along the Persian Gulf, the strategic value of · which has again attained international importance in 1899.

With his army he entered the inhospitable territory of Gedrosia (Beluchistan), and in the spring of 325 reached Susa, whence he proceeded by Ecbatana to Babylon, where in the midst of far-reaching military and political schemes he died 323 B.C.

Every young student is excited by the very name of the famous Macedonian, and a consideration of his military career is one of the most popular of geographical studies. His strategic peregrinations can easily be followed on the map by the aid of the following summary :—

He entered Asia by crossing the Hellespont between Sestos and Abydos. He then moved on to Granicus (scene of a great battle), Sardis, Miletus, and Halicarnassus (long sieges), along the sea-coast by

Phaselis, thence northward to Gordium and Ancyra, returning again to the coast of Tarsus, after passing the Cilician Gates, Issus (great victory), Tyre, Gaza, Pelusium, Memphis, Temple of Jupiter Ammon, Paraetonium, Alexandria, back to Memphis, and Syria, to Damascus, Thapsacus, Plains of Gaugamela, Arbela (great victory), Opis, Babylon, Susa, Persian Gates, Persepolis, thence northward to Ecbatana, Ragae, Caspian Gates, Hecatompylos, S. of Caspian Sea, Zadracarta, Susia (Meshed), Artacoana, Alexandria in Ariis (Herat), Phra, Kandahar, the Bamian Pass, or Panjshir River, Aornus, Baktra, the passage of the Oxus at the modern Kylil, Maracanda (Samarcand), Cyropolis on the Jaxartes, and back to Baktria. He then conducted an expedition to Bokhara, making the passage of the Indian Caucasus. He marched from the Choaspes River (Kunar) by the Suastes (Swat) to Attock, Taxila, Lahore, Amritsar, and the Beas, back by Mong, Tabumba, Multan, the Gedrosian desert, Karmania, Persia, Susa, and Babylon.

When Hannibal marched from Spain to the south of Italy (218—216 B.C.), his route was by Carthago Nova, Saguntum, the passage of the Ebro, Aquae Calidae, Roda, the passage of the Pyrenees, Portus Veneris, Narbo Martius (Narbonne), the passage of the Rhône near Avenio (Avignon), where he had a skirmish with Roman cavalry, passage of the Druentius, passage of the Isara, passage of the Alps—at the Little St Bernard, according to Polybius and Justice Bowen, by the Great St Bernard or Mont Genèvre according to others—Augusta Taurinorum (Turin), passage of the

Ticinus north of the Po (great battle against Scipio), passage of the Po, passage of the Trebbia south of the Po (great battle against Sempronius), across the Apennines, Parma, the marshes of the Arno, Faesulae (Fiesole), Arretium, Cortona, Lake Trasimene (great battle against Flaminius) Perusia, Spoletium, Asculum, Hadria, Teanum, Luceria, Cannæ (great battle against Aemilius Paulus and Varro in 216).

It is not without reason that so universal and vivid a remembrance of the Punic Wars has dwelt in the mind of man. They formed no mere series of battles to determine the lot of two cities or two empires, but a mighty struggle on the event of which depended the fate of two races of mankind—whether the dominion of the world should belong to the Indo-Germanic or to the Semitic family of nations.

The march of Nero against Hasdrubal[1], which resulted in the battle of Metaurus, is one of the finest strategical movements on record. Objections have been made that he ran a great risk of Hannibal discovering his absence and following in his rear. This is true, but Hannibal marching through a hostile country could not have gone very fast with the army of Apulia, which Nero had left, hanging on his rear. The distance, 270 miles, which by dint of the assistance of the entire population Nero accomplished in seven days, would under most favourable circumstances have taken Hannibal fourteen, or more probably twenty. Nero's calculations were based on the suddenness of his appearance in the north, and probably on confidence in his own military genius. He hoped to destroy Hasdrubal by

---

[1] See MacDougall's *Campaigns of Hannibal.*

one blow, sudden and decisive, and to return to Apulia in time to oppose Hannibal. War is a game of chance, and the general who risks nothing will gain nothing; his business is to reduce the risks to a minimum. Nero had time on his side, and time is a more valuable ally than any other. He took every possible precaution, particularly as regards secrecy, even keeping his own soldiers ignorant of their destination. His march is as perfect an example as can be afforded of the advantage of interior lines of operation. The obstacles which existed to the junction of the two brothers were created by the fact that they were operating on exterior lines. The obstacles themselves were the numerous armies interposed between them, and the consequent impossibility of concert.

There is something in Hasdrubal's conduct which is difficult to understand. If he was advancing confidently to attack 40,000 men, it does not clearly appear why he so suddenly changed his resolution. It is supposed that this was due to the knowledge that Nero was in the hostile camp, and the belief that therefore some disaster must have happened to his brother. Hasdrubal could only know of the arrival of Roman reinforcements either from the report of spies or his own observations. If from spies, they would certainly tell him that the Consul Nero had arrived, but they would also tell him of the very small force by which he was accompanied, which would show that he had left his army to watch Hannibal in the south, and dispel the idea of any great disaster to his brother.

On more than one occasion Hannibal violated the ordinary rules of war by placing himself in situations

which to men of less transcendent ability would have
led to ruin. But he measured correctly the capacity of
his adversaries and his own, and that which in another
would have been rashness, was in him only the fruit
of the most deliberate and just calculation. In this
respect he resembles Alexander, and indeed all great
generals. Alexander commenced the conquest of Asia
Minor with a force little superior to that with which
Hannibal descended from the Alps. He manifested
the same ability in creating a base of operations and
acquiring allies, or rather subjects, whom his policy
retained faithful to him. As instances of his contempt
of mere rules, Alexander fought the battle of Issus with
a narrow pass behind him, and the· army of Darius
interposed between him and his natural line of retreat.
Again, he fought the battle of Arbela having the Tigris,
the Euphrates, and the desert in his rear, in the heart of
an enemy's country, and having no depôt nearer than
Tyre.

Before the time of Cæsar the Romans had some
fighting in Gaul. An outrage upon the Roman com-
missioners sent to Aegytna[1] in response to an appeal
from the Phocaean colony at Marseilles, led to the first
crushing defeat of the Ligurians at the hands of the
Romans, and indirectly to the founding of the first
transalpine colony at Narbonne. From this base a
series of minor campaigns won Provence for the Roman
Government. But before the Romans were safely in the
saddle all Italy was convulsed with the news of the
approach of a vast horde of German barbarians called

---

[1] Identified with Cannes by Mr Hall in his *Romans in the Riviera*.

the Cimbri and Teutones, before whom five Roman armies in succession had gone down.

These tribes came from the Cimbric Chersonese, and were probably closely related to the Angli and the Saxones. They left their homes a little more than a century before our era. After ravaging Gaul, they defeated the Romans at Arausio (Orange) in 105 B.C., the total loss of the Romans being 120,000.

The Cimbri and Teutones are said to have amounted to 300,000 warriors, with their families and servants, probably not less than a million and a half of souls. The question of their commissariat is a problem for all ages. They had not come, like the Gauls, to swoop down upon Italy, and return home with their spoil, but were rather armed immigrants, bringing with them thousands of tented waggons, with their wives and children and all their possessions, to settle wherever they could find unoccupied land. After a sojourn in Spain they resolved to invade Italy and made the fatal mistake of dividing their forces. Marius, in his entrenched camp at the mouth of the Rhône, must have seen a picturesque sight as he watched the host of yellow-haired giants led by white-robed priestesses, with the sacrificial knife dangling from their girdles, defile before him. Of the momentous battle of Aquae Sextiae, which should rank high among the decisive battles of history, Mr Hall gives a spirited account, having carefully examined the battle-field at Pourrières (Campi Putridi). Here the Teutones were annihilated 102 B.C. In the following year the Cimbri, who had invaded Italy by the Brenner Pass, were destroyed by Marius after a gallant fight at Vercellae.

Cæsar's first command was when, at the age of forty, he went as proprætor to Spain in 61 B.C. and fought in Galicia and Portugal. Three years later, with four legions, he began his campaigns in Gaul, which lasted for nine years, crossing from Italy by Mont Genèvre in 58 B.C. At Autun he defeated the Helvetii, and he drove Ariovistus and his Germans across the Rhine near Mülhausen. In 57 B.C. he defeated the Belgae north of the Somme, and the Nervii by the banks of the Scheldt and the Meuse. In the year 56, while he was in Italy, a great rising took place in the north-west of Gaul. He hurried back to his command, and then conducted a great combined naval and military campaign in Brittany against the Veneti, who were the chief maritime power in the neighbourhood of the Channel, and put an end to their supremacy. The details of this expedition would be worthy of study by modern strategists. He failed against the Morini, who had a strong position, probably between Bologne and Dunkirk. He then turned on the Tencteri, who dwelt near Bonn, and the Usipetes, who were near Düsseldorf, and destroyed them on the Gallic side of the Rhine, afterwards crossing the river near Cologne and resting for eighteen days on the eastern bank. In the same year he failed in Britain. In 54 B.C. he failed again in Britain, although he managed to cross the Thames. In this year there was a series of risings in Gaul. The camp at Aduatica (Tongres on the Meuse) was attacked and the troops cut to pieces by the Eburones of Belgium. The Nervii were ripe for revolt, but Cæsar arrived in time to check them. In 53 he exterminated the Eburones. The year 52 B.C. witnessed a universal and final rising of all the

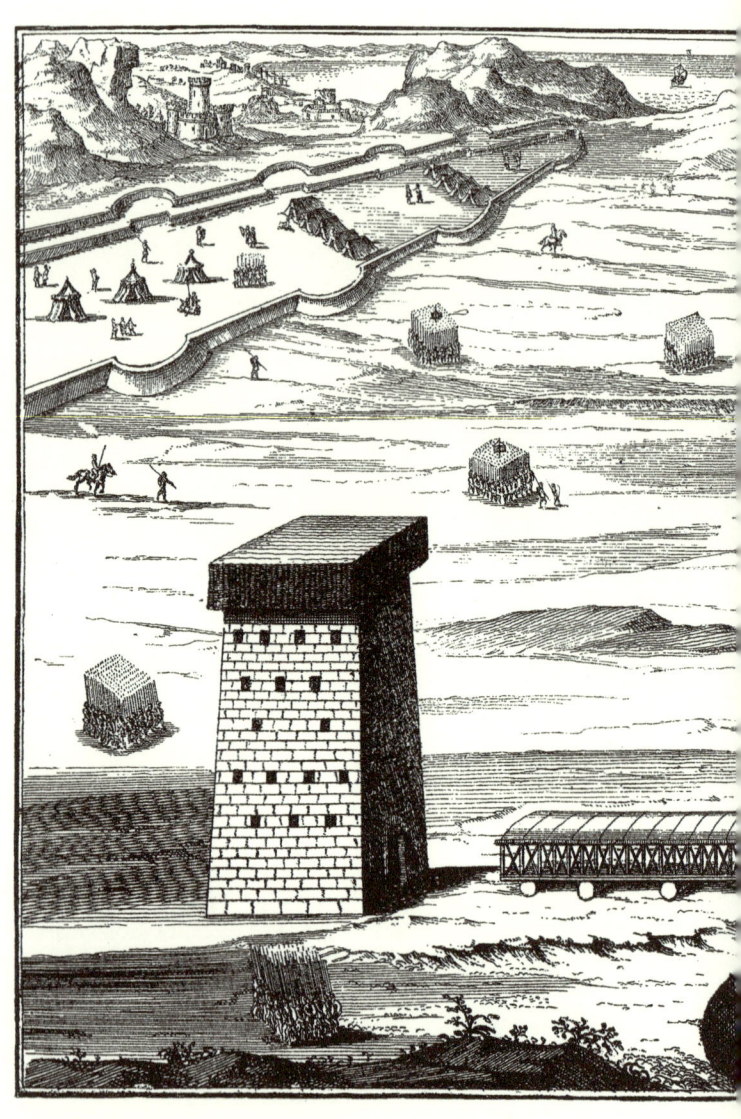

SIEGE OF MARSEILLES BY CÆSAR, B.C. 49.

Gallic tribes against the Romans, the revolt being led by the illustrious Vercingetorix. The fate of Gaul was decided between the Arançon and the Auron. On the latter river Avaricum (Bourges) was sacked and its population put to the sword; at Gergovia (Gergoie near the Allier) Cæsar was defeated, and tribes hitherto loyal joined their national leader, Vercingetorix, who collected 80,000 foot and 25,000 horse at the hill-fort of Alesia (Alise) in the territory of the Mandubii (Côte d'Or).

After a celebrated siege, in which swarms of Gauls bent on relieving their chief frequently beleaguered Cæsar, the provisions of Alesia were exhausted and Vercingetorix surrendered. The last stand of the Gauls was made at Uxellodunum (Puy d'Issolu). Gaul became Roman in language, habits, and law, and in 51 B.C. was completely pacified. Civil strife, however, now broke out in Italy, and the Pompeian party aimed at Cæsar's ruin. He led a force into northern Italy, or rather Cisalpine Gaul, and though he had only one legion at Ravenna, in 49 B.C. he crossed the Rubicon, which divided his province of Cisalpine Gaul from Italy. This was a distinct act of war. Instead of marching straight on Rome he turned it by the east, by Corfinium and Aquila, while Pompey retired to Brundusium, and he only entered Rome when he was master of Italy. Before Cæsar could completely invest Brundusium Pompey had escaped to Dyrrachium (Durazzo). His next campaign was in Spain, where he defeated the lieutenants of Pompey, Petreius and Afranius, at Ilerda in Catalonia, and on his return to Italy he took Marseilles. He afterwards embarked for Greece, not being able to force Pompey's lines near Dyrrachium—then the chief point of departure for a traveller in the East—and retiring into Thessaly was followed by Pompey, who reluctantly abandoned his base at Dyrrachium and was defeated at Pharsalia 48 B.C.

Cæsar then passed into Egypt, but his rival was no more. After the Alexandrian War he was hurried from the charms of Cleopatra to Zela in Pontus, where he defeated Pharnaces, and whence he sent his alliterative despatch "Veni, vidi, vici;" which has however been surpassed in brevity by Sir Charles Napier's punning announcement of the conquest of Scinde—" *Peccavi.*"

From Egypt he hurried into Mauretania and defeated Juba, causing the younger Cato to commit suicide at Utica (Bou Shater). After a triumph in Rome he was obliged to hurry across the Pyrenees to defeat Labienus and the sons of Pompey at Munda in the extremity of Spain, between the Sierra Nevada and the sea. He went back to Rome, master of the world, to be assassinated in his 56th year.

Napoleon, in remarking on the campaigns of Cæsar, says:—" Cæsar's principles were the same as those of Alexander and Hannibal—to keep his forces united ; not to be vulnerable in more places than absolutely necessary; to throw himself rapidly on important points; to employ largely moral means, viz. the reputation of his arms, the fear which he inspired, and politic measures calculated to preserve the attachment of his allies and the submission of his conquered provinces."

The principal westward and southward invasions in the 4th and 5th centuries A.D. were conducted by the following races :—

The *Alans* or *Alani*, a tribe who came from the neighbourhood of the Caspian Sea, where they were defeated and dislodged, or rather pushed forward, by the Huns in 375. They migrated in a straight route, due west, through southern Russia, Roumania, Hungary, Austria, and Germany to France, and turning southward crossed the Pyrenees into Spain, where they were finally defeated and submerged by the Visigoths, in 418. They are supposed to have been associated with the Vandals and the Suevi.

The *Huns*, a Mongolian tribe[1], perhaps originally from northern China, replaced the Alans on the borders of the Caspian, defeated them in 375, subjected many minor tribes, and conquered practically the whole of Middle Europe. They were defeated in 451 at Châlons by a combined army of Romans, Franks, and Visigoths, but were yet able to invade Italy in the following year. They disappear from history soon after Attila's death (453).

The *Goths*, divided into three principal tribes, Visigoths, Ostrogoths, and Gepides, are of the Teutonic race, and came originally from southern Scandinavia. The *Visigoths* established an empire in the present Moldavia and invaded the Balkan Peninsula frequently. Under Alaric they invaded Italy. Finally they came to France (417) and Spain, where they beat the Alans and Suevi, and established an empire that lasted until the advent of the Arabs (712). The word Visigoth is a corruption of the German " Westgothes," *i.e.* Western Goths. The *Ostro-*

---

[1] Prof. A. H. Keane regards the Huns as "a heterogeneous collection of Mongol, Tungus, Turki, and perhaps even Finnish hordes under a Mongol military caste." *Man: Past and Present*, p. 305.

goths appeared in Bulgaria about 460, conquered Italy in 493, established there a great empire under Theodoric, and were defeated and wiped out by the Eastern Romans in 553. Ostrogoth is a corruption of the German " Ostgothen," *i.e.* Eastern Goths. The *Gepides* settled in Wallachia, and were defeated and superseded by the Lombards in 560.

The *Lombards* or Longobardi, a Teutonic tribe, are said to have come from the Baltic coast (Mecklenburg and Pomerania). They migrated southwards, and after defeating the Gepides in Wallachia invaded Italy in 568, and established there an empire which lasted until 774, when it was destroyed by Charlemagne.

The *Vandals*, also a Teutonic tribe, came from the Baltic (the present provinces of West Prussia, and parts of Pomerania and Brandenburg). They migrated southwards, first into France, then through Spain into north-west Africa. By the conquest of Carthage in 439 they established there an empire which soon grew into a maritime Power of the first rank, under the great chief Genseric. In 455 the latter invaded Italy, and conquered and sacked Rome, returning afterwards to Africa. The Vandal Empire was destroyed by the Roman general Belisarius in 533.

The *Burgundians*, another Teutonic tribe from the eastern Baltic, settled in eastern France about 413, and were subjected by the Franks in 536.

The *Suevi*, also from the eastern Baltic, invaded Spain in company with the Alans (409) and established an empire in the present Portugal and north-western Spain, which was destroyed by the Visigoths in 582.

The *Franks*, a confederation of German tribes, located

between the Rhine and the Weser, crossed the former about 420, and established the empire of France, which has practically lasted to this day.

The *Anglo-Saxons*, from the mouth of the Elbe and Weser, invaded England in 465, and have remained there up to this day.

The *Saracens* (*Arabs*) invaded the whole of North Africa from Arabia between 630 and 700, crossed the Straits of Gibraltar, and destroyed the old Visigoth Empire in Spain in 712. The Arab Empire of Spain lasted 780 years ; Granada, the last Arab stronghold, falling in 1492.

The *Magyars*, a Turanian tribe, successors of the kindred Avars, appeared in Europe in 804, having come from the Ural districts. Their westward migration was checked at Merseburg in 933 by the Emperor Henry I., and again in 955 on the Lech by Otto the Great. They settled finally in Hungary, which they inhabit at the present day.

To the Magyars in the westward march of races succeeded the *Osmanlis* or Turks, who appeared in Europe in 1353 and completed the conquest of the Balkan peninsula in 1453 by the taking of Constantinople. The westward migration of the Turks was brought to a stop at Vienna in 1683, and they were then gradually pushed back to the Balkan peninsula. The displacement of the Turks has not yet ceased.

The territory north and south of the Caspian Sea, Smyrna, the Bosporus, Bokhara, and the Oxus and Jaxartes, the passes of Central Asia, Anatolia, Bactria, Badakshan, the feeble defences of China—outside the range of practical politics and ignored in every European capital for 440 years after the death of Tamerlane—are now eagerly discussed in every Foreign Office and every intelligence-department from London to Moscow. But instead of the cavalry of Tartary surging in myriads around the eastern boundaries of Europe and the shores of the Levant, in our generation isolated detachments of Europeans, borne in ships mightier than all the naval resources of Constantinople at their best, and armed with weapons infinitely more destructive than Greek fire, are now planting themselves on the Asiatic coasts.

A geographical sketch of the marches of Alaric, Attila, Jenghis Khan, Timurlane, and Bonaparte will suffice to justify the proposition that strategy is a simple science, easily comprehended by any intelligent man who applies his mind to it, whatever the deficiencies of his early education ; that no profound knowledge even of geographical details is necessary for successful invasion, that the main lines of road, railways, and strategic marches are always the same, and that the principles of the art of war are unchangeable.

Compared with later invaders of the Western and Eastern Empires Alaric was a mild and chivalrous conqueror. He could be bought off, and could be cajoled ; moreover as one of the new converts to, or admirers of Christianity, he treated the shrines of Rome with respect, and in all his dealings was far more honest than the treacherous and scheming cowards and assassins who governed the Empire of the Cæsars in its decline. But he was the first to show the Barbarians the road to Rome, and to expose the political, moral, and military decay of the Italians, who had not only lost the courage and hardihood of soldiers, but dared not engage in any military operations without the aid of warriors from the Rhine and the Danube. St Augustine preferred the Goths to the Romans, and declared that Alaric was an instrument of God's justice against a city which had become the mother of every error and every vice.

Descended from one of the noblest families of his nation, we first find Alaric employed by Theodosius against the Huns in 395 A.D. He afterwards engaged as a mercenary in the civil wars, but neither he nor his people were satisfied with Thrace as a recompense. He soon ravaged Macedonia and Thessaly, destroying in his Christian zeal all the works of paganism, from the Vale of Tempe to the Morea. Stilicho, one of the best of generals, for a while checked his career, but he was soon master of all Illyria, and therefore of the mountains which intervene between the Danube and north-eastern Italy. He started for the invasion of Italy 400 A.D., sacked Aquileia, took Milan, was again checked by Stilicho, but appeared before the capital, marching by the Via Flaminia, in 408. He was bought off a second time, but the feeble Honorius could neither destroy the Goths nor act fairly towards them, and Alaric again marched on Rome and captured it, in 410, and for three days the Romans witnessed helplessly the utter destruction of all the riches which had been gathered together during centuries of triumph. Fearing that his soldiers might be corrupted by a long stay in Rome he hurried from it, ravaged Campania, Apulia, and Calabria, and was preparing an expedition to Sicily when he died at Cosenza, and was buried with his treasures in the bed of the Busentinus. His soldiers killed all the captives who had diverted the course of the river and prepared his tomb.

Attila, son of Mandras, when he became king of the Huns, whose chiefs had divided up Hungary and Scythia, shared his power with a brother, Bleda, and levied tribute on the weak emperor Theodosius II. Having gained the confidence of a nation of warriors, and pretending to have secured the sword of their divinity, he got rid of his brother, and soon his sway was acknowledged by all the martial races who longed to be led once more to the spoil of southern Europe and western Asia. Remembering the days of Alaric, Vandals, Ostrogoths, Gepides, and some Franks served under his banner; nor did the fact that members of the same tribe were in Roman pay and ready to resist him affect his plans. The Hun aimed at universal empire, though not for luxury and licentious enjoyment. He never had a palace, he erected no temples, no monuments such as hand down to posterity the memory of the Central Asian conquerors: his head-quarters were a simple cabin by the banks of the Danube near Komorn, surrounded with captured trophies. Collecting together some 700,000 hardy soldiers, he proposed to spoil Persia, having heard rumours of its ancient opulence. A long march around the north of the Black Sea brought him into Armenia, where he was thoroughly defeated. In truth his first adventure was a mistake, whether strategic success or the probability of booty be considered. His next campaign was a very profitable and much easier enterprise. Having mastered Illyria, his followers laid waste all the Balkan peninsula from the Adriatic to the Black Sea, and would have taken Constantinople, only none of his chiefs understood the art of conducting sieges. During this raid he destroyed seventy flourishing cities. Before invading Italy he led his hordes through Franconia and passed the Rhine. The terrified Gauls fled into the forests and mountains, and he reached Orléans with impunity. Here the inhabitants stood a siege, and as the Romans under Aetius and the Goths under Theodoric were marching to raise it, he feared for his line of communication and retired to Châlons-sur-Marne, where occurred a justly celebrated engagement—one of the fifteen decisive battles of the world. The Gothic leader was killed, but by a skilful use of the reserves his son Thorismund won the day. Attila fortified his camp and prepared to destroy his booty in case of another defeat next day, but Aetius did not attack, and he retired across the Rhine. His retreats under the most desperate conditions were models of this branch of the art of war, and enhanced rather than diminished his reputation. According to some historians 160,000 dead covered the plains of Champagne. But a similar number of Napoleon's conscripts in 1813 were left by the banks of the Elbe, and yet their sovereign never won more glory than in 1814 in the theatre of Attila's reverse.

As the champion of Honoria, sister of the emperor Valentinian, who wished to marry him, Attila next invaded Italy, upon which he laid the scourge of God with a heavy hand. Aquileia, Padua, Vicenza, Verona were soon in flames; the fertile fields of Lombardy were wasted; the terrified inhabitants of both banks of the Po fled into the Alps and the Apennines, or to the lagoons of the Adriatic, where the fugitives from Venetia founded Venice. Milan was taken, and an artist was

employed to celebrate his successes, even as when Napoleon conquered the same city he erected his own statue
among the effigies of Adam and the saints and prophets
on the pinnacles of the cathedral.

Why he did not storm Rome is hard to say. It is
not likely that Saints Peter and Paul, as Raphael has
pretended, would have protected a city which was the
mistress of all the vices. Certainly Pope Leo risked his
life to plead for his flock ; but Attila would have paid
little regard to the sanctity of an ambassador unless it
suited his policy. A large sum of money was a more
powerful intercessor, and the prudent Hun was always
cautious about his rearward line. It was a far cry from
the Tiber to the Waag, and if foiled before Rome his
troops would perish for want of food, for northern Italy
was as desolate as was the road from Adrianople to the
Danube in 1878.

Attila tried another campaign in Gaul, but was
beaten. In his simple home he forgot his championship
of the daughter of the Emperor. He found among the
daughters of barbarism a far more fascinating temptress,
and the night of his nuptials with Ildico delivered
humanity from one of its most relentless persecutors.
His empire passed away into obscurity ; even Hungary
does not derive its name from his followers, though it
does from his race.

CHAPTER XIII.

HISTORIC LINES OF INVASION (*continued*).

ZENGHIS KHAN, or Jenghis (son of Pisoukai, chief of a Mongol horde), and his descendant, Timurlane or Timur-Beg, the former born in 1163 and the latter in 1336, emerged from Central Asia between the Oxus and the Jaxartes, to the horror of all men who dwelt in the vast territories extending from Poland to Peking, and from the Hellespont to Delhi. To follow in detail the marches of their ever-victorious armies would be to give a complete topography of all the passes and roads to be found in these regions, and to describe the siege of every city which contained enough booty to tempt the avarice of their followers. No civilization could cope with their barbarism; no general of other fighting races, Moslem or Christian, Tatar or European, could match their military genius, the rush of their cavalry, and the skill of their tactics. Not even the dauntless Ottoman Turks, under their brilliant leader, the Sultan Bajazet, could arrest their progress. But their empires depended on themselves and broke up immediately after their deaths.

The career of Jenghis for twenty-six years was one continuous series of triumphs and conquests, atrocities and massacres, destruction of cities and laying waste of

rich provinces, which have never since recovered a tithe
of their former prosperity.   All the works of art and invaluable libraries in the numerous and flourishing cities
between the Oxus and Euphrates were broken to pieces
or burnt.   Any resistance to the will of this well-educated
savage, or of his extremely able, brave, and cultivated
sons and generals, meant annihilation.   His motto was,
" The only way to make peace with an enemy is to
destroy him."   From his head-quarters at Karakorum
he marched his legions—numerous as any of the four
German armies that conquered France—south, west, and
east, backwards and forwards.   Having consolidated his
authority over his own race, he began his career of
foreign conquest by taking the territory of the Uigurs.
In 1209 he passed the Great Wall, and in 1215
sacked Peking, and burnt it for a month.   In 1218 he
was in Turkestan, fighting a drawn battle against the
Kharismians.   In 1220 Bokhara and Samarkand were
stormed, sacked, and burnt; yet their 300,000 victims
and the disappearance of the greatest library in Asia
did not satiate the destructive fury of the Mongol chief.
Next year Balkh, which had kept alive some of ,the
glories of the old Persian satraps, was in flames.   The
men, women, children, and animals of Bamian were
slaughtered to avenge the death of one of his grandsons.
In a few weeks Khorasan was overrun.   Herat held out
for six days, and the horrors which followed its capture
sent a thrill through even the Mahommedan peoples
of the East, well accustomed to traditions of brutality on
the part of kings and conquerors and priests.   Derbend
was passed, and Russia and its defenders driven to the
Dnieper in 1223.   Returning to Central Asia, Jenghis

was with difficulty persuaded not to carry out his design
of annihilating all the agricultural and manufacturing
population of China, so as to give more space for a
nomad race of horsemen.   He traversed the Gobi desert
in 1226.   He then besieged and took Nirghi, the capital
of Tangut, and destroyed 90 per cent. of the people of
this hitherto prosperous and powerful state.   This was
his last feat ; he died aged 66, and was buried in
Tangut, having become absolute master of the territory
from Kiev to Peking, an extent of nearly 4,000 miles,
and having built up his empire at the cost of at least
5,000,000 lives.   Some of his descendants are still princes
in Turkestan.   Another ruled in the Crimea till Catherine
of Russia annexed it in 1783 ; and Holagou and Kublai,
his grandsons, won high places in history for themselves.

But none of the Tatar or Mongol race was as celebrated as the lame Timur-Beg or Timurlane.   He was
born near Samarkand in 1336, and after the best education that his philosophical father Fargai could give to a
young Mohammedan chief, in the way of military exercises, horsemanship, the chase, and the Koran, he soon
began to play a leading part in the disputes which
agitated every part of these realms, and had only been
restrained by Jenghis Khan himself and his immediate
successors.   Young Timur led a romantic and wandering
life of adventure for some time before settling down to his
regular career of conqueror and slaughterer.   He began
with delivering Samarkand.   In a battle near Siestan
he was so wounded as to be maimed and lamed for life.
He drove the Khan of Kashgar out of Transoxiana.
He took Balkh, the fortress of his brother-in-law Hassan,
who was killed and his sons burned in the citadel.   He

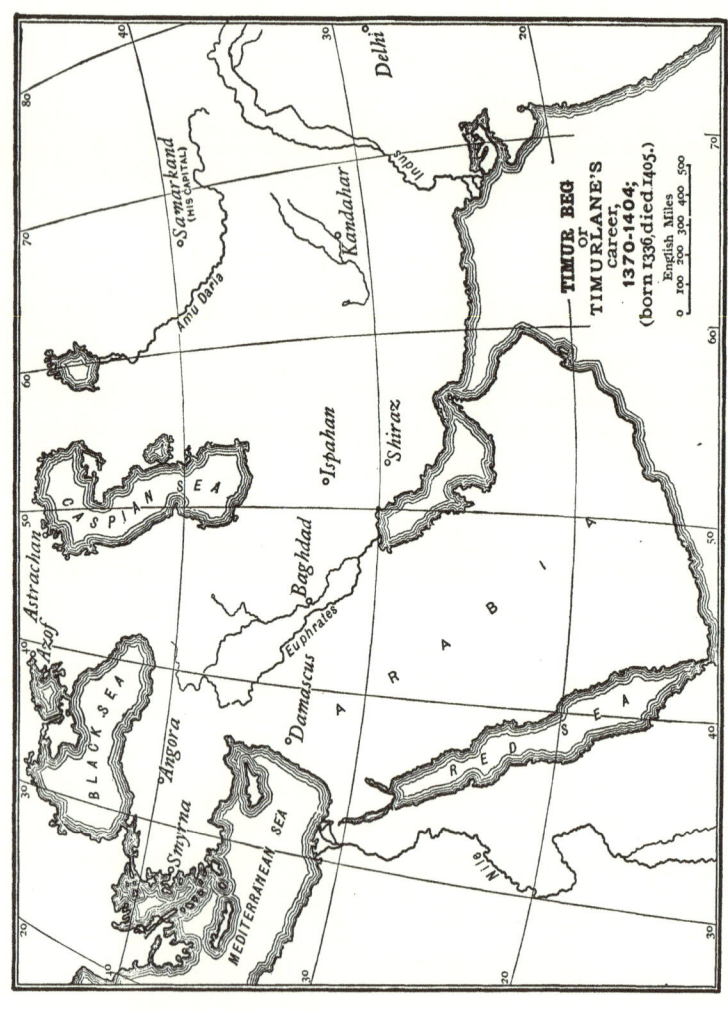

TIMUR BEG
OR
TIMURLANE'S
career,
1370-1404;
(born 1336, died 1405.)

English Miles
0   100 200 300 400 500

was enthroned emperor of Djagatai in 1370, and fixed his capital at Samarkand, which he embellished with mosques, palaces, and hospitals, and made a brilliant centre of literature and the arts. He passed the Jaxartes and made himself master of Kashgar, and conquered and destroyed the territory of the Kharismians in 1371. He next attacked Khorasan, stormed Tashkend and massacred its people, and took Herat, which he plundered to the very gates. He began here his curious and horrid custom of erecting monuments of heads; and after the capture of Setzwar he murdered all the inhabitants, except 2,000, whom he used to make towers, heaping them alive one over the other, and fastening together the monstrous edifice with bricks and mortar.

In the same year, in person or by his warriors, he covered Siestan, Afghanistan, and Khotan with ruins and corpses. In 1384 he utterly destroyed Asterabad. He rested for a few months in his capital, and starting in 1386 for Georgia, took Kars and Tiflis. Year after year the Tatars perpetrated atrocities in this Christian kingdom, but they were impartial, for they were even more cruel to the followers of their prophet in Persia and in Syria, and overran Armenia, then ruled by Turkomans. In 1387 Timurlane reached Ispahan, in Persia. As some of his soldiers perished in a chance riot in the city, he ordered 66,000 heads to be carried to the ramparts of the town, where they were registered and built up into towers. In the following year he drove any of his enemies who remained in eastern Tartary beyond the Irtish. In 1390 he again depopulated Kharismia, destroyed the capital, sowed barley over the soil, and carried off the people to Samarkand. As his next task he resolved to conquer the Tatar empire of Kiptchak, which had been founded by Jenghis and stretched from the Dnieper to about 100 miles north of Tashkend. He was now fighting a brave Moslem adversary, led by the able Toktamisch. Had Timurlane not undertaken this desperate and well-nigh fatal enterprise, the present empire of the Tzars could never have been developed, and the crescent of Islam would have superseded the cross in all south-eastern Europe. After a terrible march lasting four months—in which his men, having lost touch with regular supplies almost as completely as those of Napoleon in 1812, lived on herbs and the eggs of wild-fowl, helped by the products of the chase, like Wellington's army after Talavera—he routed Toktamisch between the Yaik and the Volga, and rested on the banks of this river for a month. He seated himself at Serai on the throne of the Khans, and returned to his capital. He left it again in 1392, and next year having ravaged Kurdistan, he advanced against Baghdad, which he took, as also Basra and Mussul. As Toktamisch again proved troublesome, he massed 400,000 men for the ruin of what is now southern Russia. He traversed safely the tremendous defile of Dariel, half way between the Black and Caspian Seas, and crossed the Terek. Some say he sacked Moscow, but of this there is a doubt. Ordering his grandson to devastate the western provinces of Russia and Poland, he returned, razed Serai and Astrakhan to the ground, and reduced their citizens to slavery, carrying off all their wealth. But as was the case 417 years later, the climate and the solitudes of Russia took vengeance on the invader, and no amount of booty could compensate the nomad chief for his terrible losses in men and money. After being away for five years he returned to Samarkand, still further wasting what was left in Georgia on his route.

Exhausted as was his army he had no difficulty in recruiting its ranks, especially as the loot of Hindustan was next offered to the children of the desert. His cavalry comprised the best horsemen of Asia, in number 80,000, and his way over the Hindu Kush and the mountains from Kabul to Herat was explored and cleared by an advance guard under Mohammed Djihanghir, who had long lowered over the north-west borders of India. This march was one of the most marvellous in history, and is well worthy of careful study in full detail by every British strategist. It was the model for Skobeleff's proposed raid on our Empire. In the midst of snow-clad passes Timurlane found his progress delayed by pagans who claimed the Macedonian soldiers as their progenitors, and whose descendants between the Swat and the Panjshir are now under British rule. The climate and rocks and scarcity of forage were fatal to a large proportion of his horses, but after six months he reached the Indus. From this ancient base of operations he soon mastered Delhi, but as he found his movements checked by myriads of captives and followers, he massacred 100,000 of them before committing himself to a decisive battle with the Sultan Mohammed. Similar cruel deeds, from similar strategic reasons, were committed in Syria by Bonaparte, and the negro followers of Sherman were left on the wrong side of a river during the latter's march to Savannah. The sack of Delhi in 1299 and the marvellous stores of wealth which were carried back to Tartary have ever since been part of the folk-lore of all the tribes from the Altai Mountains to the Caucasus. They would gladly welcome the cruelty of another Timur who would conduct them either around or over the " Roof of the World " for a year's riot by the banks of the Ganges—a far more profitable avocation than seeking scanty pasture for their flocks and herds on their own side of the mountains.

Having exterminated some myriads of Indian idolaters on both banks of their sacred stream, Timur returned home to pay homage to his prophet by the erection of a magnificent mosque in 1399. Georgia gave him some trouble again, though its people strove in vain, but he had soon to turn his attention to the Ottoman Turks, who, having conquered almost all the territory of the Grecian Empire except Constantinople itself, were in no mood to submit to the dictation of a wild chieftain from Central Asia, the home of their own ancestors.

Western Christendom has of late been singularly out of touch with the true interests of the East, and denunciations of the rule of the Turks current since 1876 have had no historic foundation. In 1400 the forces of Bajazet, in Asia Minor, were even more certainly the outposts of Christian Europe than Antwerp, Ostend, and Belgium at large have been outposts of Great Britain. The effeminate and despicable rulers of the Greek Empire would have gone down without a struggle before the masterful Tatar, but the Turks challenged him, and, though defeated, did not yield without heroic efforts which made the battlefields of the campaign 1400–1404 among the most remarkable in the military annals of mankind. The scope of this book prevents details of tactics on the mightiest of fields, but the reader would

be well advised if he perused the glowing pages of
Gibbon, who is at his best when describing this war of
giants.   As usual, Timurlane took the initiative ; at
Cæsarea in Asia Minor he cut in pieces the army of the
son of the Ottoman Sultan.   He besieged and took
Siwas, and buried alive 4000 wretches who garrisoned
its citadel, but he turned aside for a short period to
conquer Syria, then governed by the Mamelukes of
Egypt.   When his soldiers were building towers of
human heads after the sack of Aleppo he discussed
questions of theology and ethics with the savants, whom
he always spared.   He took the venerable Damascus by
a ruse and burnt it.   It is a maxim of strategy not to
stretch a line of communications too far, especially if on
the flanks are competent adversaries ; thus Timurlane,
who had like Napoleon an instinctive genius for war,
was careful not to go too far into Russia in his previous
campaigns, in which respect he was more astute than
Napoleon, and with the Turks in front he also wisely
abstained from going too far south and following the
beaten Mamelukes into Egypt.   This would have led
to surrender in case of defeat, and outraged humanity
would have applauded the Mamelukes if they had supplemented the immortal tombs of the Pharaohs by
another pyramid composed of the bones of himself and
his followers.

Before marching against the Turks in Asia Minor he
turned from Syria to the Euphrates.   His tents soon
covered the environs of Baghdad.   On the 9th of July,
1401, the home of the Caliph began to experience the
terrors of obliteration.   For eight days the Tatars were
employed in massacre.   Crowds of despairing Moslems

rushed into the river to escape a doom far more terrible than that whereby, many centuries before, the Medes and Persians eclipsed the grandeur of Babylon. Mosques, colleges, hospitals, all disappeared. Not a trace was left of the monuments by which the Abbasides had hoped to win the respect or the pity of posterity. Instead, 90,000 heads were piled up in 120 towers—paltry and perishable records as compared with the hanging gardens and the palaces of Sardanapalus. The Tatars then left the valleys of the Tigris and the Euphrates, which have ever since been the synonyms of greatness buried in ruins.

If any power could have conquered our awful hero it would have been the Ottoman Turk. Timurlane took up winter quarters on the Araxes, and hurried up recruits from all parts of his empire till his army was 800,000 strong. He advanced in the spring of 1402 and laid siege to Angora. The magnanimous ruler of the Turks came to the relief of the place with 400,000 men. Then took place a tremendous battle, equal to Leipsic, far more dreadful than Borodino or Gravelotte. To condense Gibbon's narrative would be to spoil it. Among the peculiarities of this fight was the use of Greek fire on land, and the large herd of elephants which Timur had brought from India. The Turkish army was cut to pieces, Bajazet as a prisoner was brought before his invincible foe, who burst into a fit of laughter—not to insult his captive, but at the caprices of fortune, which had enabled a half-blind cripple to become master of the Eastern world. Smyrna had resisted the Turks for seven years, but in spite of the skill and bravery of the Grand Master of the Hospitallers, Philibert de Naillac, it was taken, sacked, and the people massacred to a man, and Timurlane returned satiated with success to Samarkand in 1404.

All Europe was aghast—if the Turks had joined their Asiatic co-religionists its civilization would have been doomed. Vienna, Augsburg, Paris, and Rome would have shared the fate of Aleppo, Baghdad, and Smyrna. But the Turks joined the Greeks, and both patrolled the Hellespont and the Bosporus. The wild horsemen of the steppes gazed with vain covetousness at the wealth and the glories of architecture that surrounded the Golden Horn, but they could not cross the Straits, they had no boats.

Napoleon's corps were in a like position in 1804. They saw the white cliffs of Albion, but the Channel was closed to them, and they had to transfer their eagles from Boulogne to Ulm. So Timurlane, despairing of a raid into Europe, returned to Samarkand, in the decoration of which he employed the cleverest artists and masons of Persia and Syria. The new buildings by the banks of the Oxus served to remind the exiles of the splendour of their own early homes by the Abana or the Euphrates.

To-day the desert has in turn covered the villas of the courtiers of the merciless conqueror, and chieftains from the banks of the Volga and the Don govern with an iron hand amidst the scattered columns and dilapidated walls of his Imperial Palace. "But quiet to quick spirits is a hell," and this was the bane of the great Tatar, as it was to be the bane of the great Corsican. Alexander wept because there were no more worlds to conquer. Timurlane declared "As there was only one God in Heaven, so there ought to be only one Emperor on earth," and "What is the earth with all its people to a ruler such as I ? "

Europe being out of the question he resolved to conquer China. His fiery zeal would not wait for spring; he set out from Samarkand in February, though the ground was deep with snow, and before long the cold made great gaps in the ranks of his 200,000 cavalry. "Tempted Fate will leave the loftiest star"—he got a fever and died at Otrar in 1405 at the age of 69.

A study of his career, and a simultaneous study of the newspapers and magazines of to-day show that in strategic geography there is nothing new under the sun, and almost suggests that Plato's theory of the recurrence of events at regular epochs has been justified by history.

The campaigns of sovereigns like Charlemagne, Gustavus Adolphus, or Frederick the Great; or of servants of their State like Turenne, Marlborough, and Suwarrow, bear no resemblance to the operations of "world-conquerors" either in regard to their objects or their conduct. But with the success of the French Revolutionists over the old monarchy and aristocracy a new power, as terrible as any horde of Goth or Hun, appalled Europe, and for twenty-two years war ravaged the land from Egypt to Denmark, and from Gibraltar to Moscow. Jenghis Khan's operations cost 5,000,000 lives, but 2,000,000 Frenchmen alone perished in 20 years. The "Liberty, Equality, and Fraternity" of France and its hero Napoleon cost £4,000,000,000. The Sultan Suleyman, though the terror of his time, was a benefactor of his species compared with the Republican General. The ferocity of the Tatars against fellow Moslems was mild compared with the atrocities which characterised the early excesses of the French Reign of Terror. Alaric was a polished gentleman compared with Bonaparte. The Huns had a heart, as had Cæsar; the Corsican was all mind, but a mind the most capacious in comprehension, and the quickest in apprehension of which history has left a record. We must pass over the marches of generals like Pichegru, Jourdan, Moreau, all very excellent, but relating to local rather than to general strategy, and come to Bonaparte himself. We cannot here dwell on the strenuous studies and the intrigues and difficulties of his career before, as the queller of a Parisian insurrection, he had earned a right to promotion. For our purpose it is sufficient to deal with his movements from the period (1796) in which he was in command of the Republican army in Italy. Henceforth to follow his career is to trace on the map the main strategic routes of Europe. It is always better to begin the study of a series of military operations by a general view of military geography and then to go back to the local topographical incidents and the details of battles. The strategy of the theatre of operations should always take precedence of the battle-field.

Bonaparte's first manifestoes in Italy, when he took command of the half-starved republicans of the Riviera, might well have been copied from the speeches of Attila to his Huns—"Soldiers you are naked and ill-fed. I will lead you into the most fruitful plains in the rich provinces; great cities will be in your power." He soon had separated the Austrians from the Sardinians by the fights of Montenotte and Mondovi. Eighteen days later, April 28th, 1796, the former gave up the war and their principal fortresses. The Po was crossed at Piacenza, Milan occupied, and by June Lombardy had been subdued and thoroughly well plundered by "contributions." Mantua was invested in July, and was taken in January, 1797, in spite of all the Austrian efforts to relieve it. Bonaparte now did as he pleased, independent of the Parisian government, to which he sent plunder from time to time. He also enriched his officers and men at the expense of the Italians.

Again, in February, 1797, the Papal territories were overrun, Leghorn was seized, though in a neutral state, and English property confiscated. A campaign was conducted in the continental part of Venetian territory. Venice itself was taken and a system of plunder instituted, the best pictures and statues of Italian towns being transferred to France. His next move was towards the north-west of Venetia. He drove the Archduke Charles from the Tagliamento to the Julian Alps and marched on to Leoben, while Joubert, commanding the Brenner Pass, was threatening the Austrian right by the road to Villach. This was a brilliant campaign, and proved that the historic and geographical researches of the young warrior were profound. He treated the French Corps Legislatif even more strenuously than Cromwell treated the Long Parliament. All feeble or factious republican Parliaments are at the mercy of the first successful and popular general who pleases to abolish them.

A lull followed the Peace of Campo Formio, 1797, but during this year the plan for invading England failed, and the navies of Spain and Holland allied to France were destroyed. Bonaparte's expeditions to Egypt and Syria, 1798, were based on colonial and maritime considerations relating to England alone, and in no way concerned European policy except in so far as his absence, by depriving France of her best general and army, enabled the Austrians and Russia to reconquer Italy, and in spite of the admirable strategy and tactics of Massena in Switzerland to threaten an invasion of France from the Riviera, Turin, and the Black Forest. When he returned, he got rid of all political complications and put an end to a worthless form of government by a *coup d'état*, and as First Consul set to work to drive the Austrians not only out of Italy but, with the aid of Moreau, out of southern Germany also. French troops suddenly appeared in May at all the passes of the Alps, Bonaparte's own road being by Geneva, St Bernard, Aosta, Vercelli, Milan, Piacenza, Marengo. He was in Geneva, May 9th, and after astonishing and reckless—but successful—strategical movements, he was back in Paris before the end of June. But it was Hohenlinden, rather than Marengo, which caused the Austrians to make peace at Luneville, 1801. His naval preparations against England and the organization of his army at Boulogne occupied his attention for a few years. But the invasion, probably seriously intended, was utterly impossible, and in 1805 his corps marched by all the leading passages of the Rhine—Augereau on the right, and Bernadotte on the left—to the Danube, around Ulm, which was taken, and to Vienna, and thence to Austerlitz.

It is strange that a man who, not only in warlike

affairs, but in every department of political life, had such a clearness of view that Marmont and others of his ablest associates regarded his prescience as almost divine, should have utterly misunderstood the strategic and commercial position of the United Kingdom. It seems that naval strategy had not been within the range of his early studies, and after he became a power in the world he had no Mahan at his elbow. Yet he was quick to observe any mistake on the part of our Admiralty, such as the temporary withdrawal of the fleet from the Mediterranean, and he managed at least as much by reason of his own astuteness as by the skill of his naval officers to deceive and elude Nelson in 1798, to carry a great expedition to Egypt, to traverse the Mediterranean with safety on his return next year, to deceive our naval authorities in 1805, and to escape our cruisers again in 1815. He also was a hearty admirer of the physical and moral excellences of our sailors. Yet all his schemes against the United Kingdom from 1803 to 1812 were based on ignorance of the true state of affairs, as for examples his armed neutrality, 1801, his invasion schemes, 1804-5, his continental system, 1806, and above all his policy as embodied in the Peace of Tilsit, 1807. It is difficult to understand his rupture with our kingdom in 1803, and the violent and unprecedented methods which led to it. "Had he remembered the teaching of his favourites Plutarch and Polybius, he could not have blundered in such a disastrous fashion."

Professor Seeley has asked, "Why he engaged in a war in which he was condemned to be so purely passive?" There is no answer available. And the next sentence of Seeley's criticism is beyond question.

" In eleven years of war Napoleon was never able to strike a single blow at England, while that enemy destroyed his fleets, conquered his colonies, and by arming all Europe against him at length brought down his power." As the sea was closed to him he began to contemplate an overland march to India, and notwithstanding all his European cares found time to negotiate with Persia. The year 1806 witnessed the wonderful campaign whereby, starting from the Main between Baireuth and Würtzburg, October 7th, he crushed the Prussians on the Saal at Jena and Auerstadt, October 14th, and reaching Berlin on the 25th, seized the Oder on the one hand and Lubeck on the other, and crushed Prussia out of an independent existence by November 7th. The fortresses of the Vistula from Warsaw to Thorn next fell. The Russians retiring from the Ukra into East Prussia were defeated February 8th at Eylau, but frost suspended operations for months. Meanwhile Dantzig was besieged, and fell in May. In June Napoleon gained such a victory at Friedland that a peace with Russia followed, the treaty being signed at Tilsit on the Niemen. The Czar became the friend of the Corsican, who for five years more was master of all Europe except the Iberian Peninsula. It seemed as if he would have found it an easier prey than it had been in the days of Hannibal, Tarif, or Peterborough. Junot took Lisbon, and the Emperor himself took Madrid in December, 1808, but the British appeared on the scene. Napoleon compelled Sir John Moore to retreat, but, as he was recalled to South Germany, the Peninsula became the grave of the reputations of his Marshals and the commencement of his own ruin.

M.   21

However, the end was not yet. In 1809, on the Danube, his corps in a few weeks carried their eagles from Ulm to Ratisbon, nor could the Archduke Charles with all his superior advantages on the Isar prevent the victories of Landshut and Eckmühl. He was at Vienna May 10th, and by July 6th the Austrian resistance was at an end. The triumph of Wagram more than compensated for the check at Aspern.

His second marriage with a Hapsburg princess conveyed no idea of domestic repose to Napoleon. He was no Alexander; his own descriptions of his endless labours while apparently he rested for two years,

> " were a school
> Which should unteach mankind the lust of reign or rule."

Not much strategy or geography, however, can be learned from his Russian campaign of 1812. The Niemen, Wilna, the Beresina, Smolensk, Borodino and Moscow showed no strategy on either side. Slaughter, scarcity, and fire made havoc during his advance; starvation, frost, and snow, during the retreat. The whole campaign was a terrible mistake, and during its course both the science and art of war were lacking. The conquering Russians suffered as much misery as the defeated French. The conflagration of Moscow was a boon to Russia as an awful warning to all future conquerors that the children of the Czar can never be subdued till the bonds of their community be shaken. France received again its emperor without his army, but in no degree crestfallen, and in 1813 on the Elbe his genius was serene against the coalition of Russians, Prussians, and Austrians. He struggled hard, but in vain, between the Elbe and the Oder. He was not beaten at Dresden, August 26th, but he was ruined at Leipsic, October 18th, the most tremendous European battle since the days of Attila. Retreating across the Main and Rhine he stood at bay in Champagne. In vain were all the efforts of the most consummate skill in war yet displayed by himself or preceding generals. After fighting a dozen battles in as many weeks he was obliged to capitulate. His residence under surveillance has immortalized the island of Elba, but he reappeared in France in 1815 and for the "Hundred Days" struck Europe with terror. His campaign in Belgium was far from a happy effort, whether strategy or tactics be considered, and yet his plan of interposing between Wellington and Blucher, driving them apart, and keeping one employed by a detaining force while he thoroughly defeated the other, was sound in principle. The 18th of June might well have been signalised by the disastrous defeat of the allies, Wellington hurrying through Brussels to his ships, and Blucher speeding towards the Rhine. The tenacity of the British at Waterloo saved Europe from this calamity, and the arrival of the Prussians on his right flank in the very crisis of the battle put an end to the career of the modern Cæsar, who—like the Roman— had fought in Belgium, Germany, Italy, Spain, and Egypt, and had been met and foiled by the British.

CHAPTER XIV.

INFLUENCE OF CLIMATE ON MILITARY OPERATIONS.

The general view of the effect of climate upon war is thus set forth by Lord Bacon :—

"Wars in ancient times seemed to move from East to West, for the Persians, Assyrians, Arabians, and Tartars were all Eastern people. It is true the Gauls were Western, but we read of only two incursions of theirs—one to Gallo-Graecia [Gallicia in Asia Minor] the other to Rome; but East to West have no certain points of heaven, and no more have the wars from East to West any certainty of observation, but North and South are fixed, and it hath seldom or never been seen that the far Southern people have invaded the Northern, but contrariwise, whereby it is manifest that the Northern tract of the world is in nature the more martial region— be it in respect of the stars of that hemisphere, or of the great continents that are upon the north, whereas the south part, for all that is known, is almost all sea (this guess has been corroborated by subsequent discovery), or, (which is most apparent) of the cold of the Northern parts, which is that, which without the aid of discipline, doth make the bodies hardiest and the courage warmest[1]."

[1] *Essay* LVIII, On the Vicissitudes of Things.

This very able analysis, considering the date of its composition, has since been accepted almost as an axiom. Yet it cannot stand the test of either ancient or modern history. In truth, the temperate zone which is intermediate between the north and the equator has produced all the greatest conquerors, from Semiramis to Alexander, from Hannibal to William of Normandy, Gustavus, and Napoleon. The majority of the Moors must also have come from the temperate zone, as certainly came all the Manchurian, Mongol, and Tatar leaders. Even the greatest patriots and liberators of history belonged to the same zone as that which gave birth to Arminius, Matthias, Henry IV., Washington, and Bolivar. It will be found that the temperate zone is the best for military as well as civil virtues, but Lord Bacon's saving clause regarding discipline applies whether to the northern and southern parts of the temperate zone, or to the world at large. A well-disciplined army, with weapons up to the highest resources of the age, and with well-organized systems of transport, will defeat an undisciplined force wherever its base, north or south. The Macedonians were north of the other Grecians, whom they routed as they pleased. The Romans were south of Helvetia and Gaul, yet they defeated the hardy and daring inhabitants of these regions. The Saracens for a period carried all before them, east and west and south and north.

Lord Bacon has made another mistake in saying that the cold of the northern parts makes the courage warmest. He probably was misled by the successes of the Visigoths, Anglo-Saxons, Scots, English, Scandinavians and Muscovites, but all these belonged to the

temperate zone.   Africa was unexplored in his time,
and therefore he had no idea of the desperate valour and
hardy enterprise of the Zulus and Matabili, or of the
strong frames and reckless heroism of the natives of the
tropical Sudan, whether the Dervish rush at Omdurman
or the Foulah cavalry charge against Arnold at Bida be
considered, nor had he any knowledge of the operations
of the Mahrattas, who went conquering and to conquer
from Poona to the Indus.   No native of Europe is
braver than a tropical African, but then the latter has
fanaticism as a substitute for discipline.   When he is
disciplined and well-armed he is capable of the greatest
things, and only armed with the spear and shield of
savagery he was able to annihilate Egyptian armies,
endanger the square at Abu Klea, and sack Khartum.

Nor is it the case even that the most northern peoples
of the temperate zone have "the hardiest bodies and the
warmest courage."   The people of Spain have been at
least as brave as the people of Holland till both ruined
themselves and their subject states by an ignoble grasping at immediate pecuniary gain.   After the Spanish
fury at Antwerp the Dutch "beggars of the sea" won a
famous reputation.   The men of Cornwall and Kent are
no whit inferior in any respect to the dwellers north of
the Highland line, and never have been so since the
days of Hengist and Horsa and of Egbert.   In the
American War the men from Texas, Georgia, Carolina,
and Virginia fought far better man for man with inferior
weapons and resources, than the men from Boston,
New York, Chicago, and Pittsburg.   But stranger than
all, the natives of southern Europe in Napoleon's army
bore the cold of Russia in 1812 better than the Russians

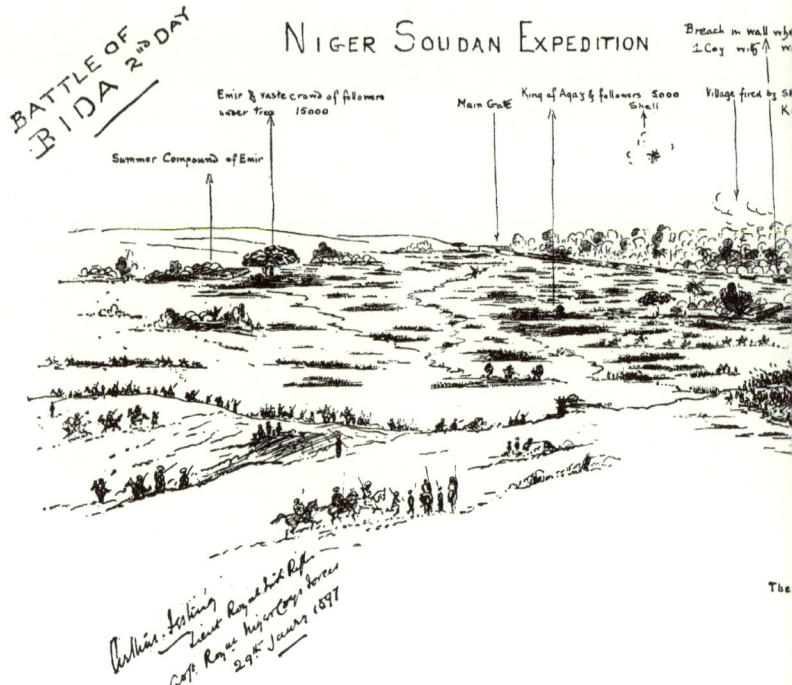

NIGER SOUDAN EXPEDITION

BATTLE OF BIDA 2ND DAY

Emir & vaste crowd of followers under tree 15000

Main Gate

King of Agay & followers 5000 Shell

Village fired by Sh K

Breach in wall whe 1 Coy with w

Summer Compound of Emir

Arthur Gosling
Lieut Royal Irish Rifle
Capt Royal Niger Coys forces
29th Janry 1897

The

SKETCH OF THE BATTLE OF BIDA, JAN. 26th, 1897.

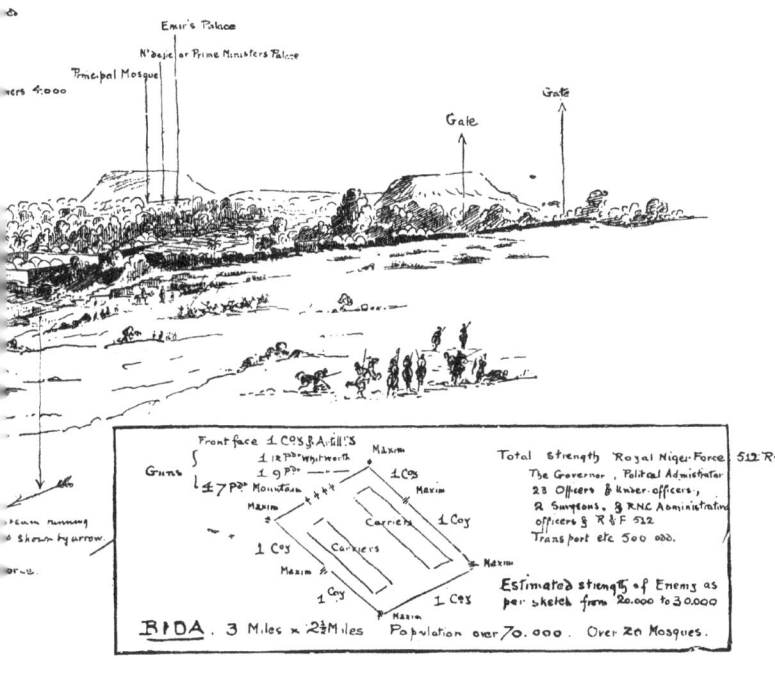

Emir's Palace

N'dage or Prime Minister's Palace

Principal Mosque

Gate

Gate

Front face 1 Co⁹ ⅜ A: till's
1 12ᴾᴰʳ Whitworth
Guns { 1 9ᴾᴰʳ — — — —
1 7 Pᵈʳ Mountain + + + + +
Maxim *
Maxim *

1 Cᵒʸ
Carriers

Maxim *
1 Coʸ

1 Coʸ

Maxim *
1 Cᵒˢ
Carriers
1 Coy

Maxim *

1 Cᵒˢ

Maxim

Total strength Royal Niger Force. 512 R⁴⁹
The Governor, Political Administrator
23 Officers & kuber. officers,
2 Surgeons, 8 RNC Administrating
officers & R & F 512
Transport etc 500 000.

Estimated strength of Enemy as
per sketch from 20.000 to 30.000

BIDA. 3 Miles × 2½ Miles Population over 70.000. Over 20 Mosques.

ers 4:000

rcam running
s shown by arrow.

orus.

themselves, and men from the basins of the Rhône and the Po displayed far more powers of endurance than men from Picardy, Belgium, and Prussia.

This interesting fact, eloquently described by Alison, is again made the basis of a considerable argument in the latest Italian treatise on Strategic Geography[1]. As compared with Turks from Asia Minor, the Russians made a very poor display of staying power when exposed to hardships in the Balkan Peninsula in 1877. Australians born under the tropic of Capricorn, and Canadians from the banks of the Saskatchewan, would together prove more than equal in physical prowess to the natives of Scandinavia and the Hungarians, whether courage, height, weight, or agility in manly exercises be considered. A bad climate will not outweigh the good of race feeling, of religion or patriotism, of delight in exercise and love of fame ; nor will a good one in any way avail licentious monarchies or corrupt democracies. The races of northern India and its north-western environs, with practically the same climate and civilization, differ enormously in courage. Compare, for example, the people of Scinde and Kashmir with the Beluchis, Pathans, and Rajputs, Hunza Nagars and Dogras.

But for many reasons, such for instance as the necessity for constant exercise and climbing, hard fare, as well as hard climate, the people of mountain districts ought to defeat the people of the plain, if equally well-disciplined and well-organized. Hence the value set upon Swiss[2], Scotch, and Swedish mercenaries in European

---

[1] Carlo Porro, *Geografia Militare*, 303. Torino, 1898.

[2] The Swiss guards of Louis XVI., as Napoleon said, if led by a resolute

wars from 1560 to 1763. To the same cause may be
traced the high esteem in which the Gurkhas are held
among the numerous races that now march under the
banners of St George. Dwellers along the slopes of
mountains resent intrusion into their valleys far more
than dwellers by great rivers or in champaign countries.
Charles the Bold, Duke of Burgundy, left his host at
Morat "for ages to remain, themselves their monument,"[1]
and the Montenegrins, Albanians, and Kurds have
never been thoroughly conquered. Hence the desperate
struggle of Schamyl and his Circassians against the
Russians, and hence our recent troubles with Waziris,
Hasaras, and Afridis.

That certain races cannot perpetuate their species in
certain countries is very well known, as for example the
British in India, but apart from this fact it seems that,
though barbarians may live in the tropics without becoming enervated, a settled civilized community in warm
climes and easy circumstances, and for a long period
free from invasion, must become utterly ruined from a

---

prince would have stopped the Revolution, but they were massacred in
vain; the Scotch were the favoured guardians of the French sovereigns
and protectors of their tombs from the days of our Henry V.

[1] With regard to the Swiss, whose present army, though small, would
certainly check any neighbouring great Power, Goldsmith has explained
their position with matchless facility :—

> "Thus every good his native wilds impart
> Impress the patriot passion in his heart.
>   *   *   *   *
> Dear is that shed to which his soul conforms,
> And dear that hill which lifts him to the storms;
> And as a child, when scaring sounds molest,
> Clings close and closer to the mother's breast,
> So the loud tempest and the whirlwind's roar
> But fix him to his native mountains more."

military point of view, even if they do not deteriorate
physically.   Under tropical skies and under repressive
conditions of religion and government, it will be admitted
that the tendency is in a marked degree to laziness.
People living by the sea ought to be smarter than inland
folk, especially if they are given to commerce like
the eastern Americans, the British, the people of the
Low Countries, the Greeks, and the modern Germans on
the Baltic and by the mouths of the Elbe and Weser.
Once they commit themselves to the waves and get the
nerves of "triple brass" which Horace ascribes to navigators, they soon become accustomed to decide quickly
and to act promptly in circumstances of danger, and
thus they acquire the very first and finest characteristic
of the military spirit, *Mens æqua in arduis.*   Moreover
their steadiness of nerve and quickness of resource lead
to a general advance in civilization.

It is true that in temperate regions intellectual
development is stimulated by conditions neither too easy
nor too hard.   Man tries to make nature his servant, and
when seasons succeed each other regularly, and by
gradations seldom sudden, his skill is developed by
taking precautions to evade the effects of great heat or
cold.   Extremes of heat and cold are causes of lethargy,
but they do not depend altogether on latitude ; there
may be very different climates in the same country,
depending on altitude, exposure to certain winds, con-
tiguity to the Gulf Stream, forests, cultivation, drainage,
radiation, and character of the soil.   Purely continental
regions are much colder than islands or peninsulas in the
same latitude.

There can be no more delusive theory than that

which lays it down that character is a function of latitude, although no doubt the love of ease most prevails in countries where one "tickles the soil with a hoe and it laughs a harvest," and in lands where little clothing or shelter in the way of permanent dwelling is required. Negroes in the West India Islands have retrogressed accordingly since the costly and premature emancipation policy of 1833. A similar fate was with difficulty fended off from the States of the defeated Southern Confederacy after 1865 by the vigour, resolution, and unscrupulous assertion of supremacy of the whites. The merciless rigour, outside the law, whereby the Galway Lynch family suppressed piracy has held the black population in awe of their former masters.

China from the Si-kiang or West river to the Amur is in the temperate zone, therefore its present condition can in no degree be ascribed to climate, and the vigorous frames of all classes of the population and their excellent work in Hong Kong, California, on the Central Pacific Railway, and in Australia, as well as their patience in face of suffering and death, their perpetual industry, and their very great skill in many an art, elegant or colossal, prove that in many respects they have not degenerated for thousands of years. Their present decay is due to political and social causes and not to climate. The false philosophy and humanitarianism of the ruling classes, wholesale corruption, rigid adherence to old custom, and—above all—contempt for the military career, have brought China to its present pass. Its future depends entirely upon the jealousies of Western nations, and is the most intricate social and political problem which has presented itself since the days of

Charlemagne.   If it be not wisely treated by our statesmen, humanity at large will suffer for centuries. Europe
would in less than four generations be the prey of other
continents if disarmament were adopted and military
training ignored.

It must further be remembered that, when the theatre
of operations is entered, nothing is of much importance
except military organization.   The inhabitants, however
well used to the climate, will perish if unprepared ; while
the invader, if provided with all requisites of clothing,
food, and drinks, will thrive.   Such was the relative
situation of the French and Germans on the Lisaine in
January, 1871.

Plains produce heroes as well as mountains ; the
fortitude of the Red Indians of the Mississippi basin, and
their gallantry in war before they were utterly broken by
the United States, are described by Cadman as being
far superior to anything conceived by the most severe of
the Spartan kings.   If there be a tithe of truth in his
story—and there seems to be no reason for doubting
him—they must have been the very bravest and most
long-suffering of men.   The ordeals which their young
men endured before being admitted to the degree of
warrior were terrible.

A generation ago students laughed at Herodotus
and his pygmies, but unless there be a conspiracy in
favour of these little folk among modern travellers there
can be no doubt that not only was he not far from the
truth, but that the forests of darkest Africa produce a
mannikin race of hereditary repute for skill in arms,
whose alliance is eagerly sought, and whose prowess is in
inverse ratio to their size.

In recent years we have learned much about the dwellers in territories within or adjacent to the Arctic Circle.   The very extremes of cold and desolation cannot eradicate valour from hardy men.   They tempt the dangers of the ocean alike between Labrador and Greenland, and between Alaska and Kamschatka.   The Tchuktchis of Bering Strait will risk themselves in heavy seas on the flimsy fabrics of their *bidarras*, and in searching for food, or for the pelts of the seaotter, display qualities of seamanship and adventure that would do credit to the British followers of Drake in the 16th century, or to the Japanese of Admiral Ito in the 19th.

Historical examples might be adduced almost indefinitely to prove that neither a favourable nor an unfavourable climate affects the military career of states, and that nations must not rely for success on any feature of their national life, on any resource except that of being a race of military men.   Yet though climate is not a determining feature in the art of war, it must, like all other natural phenomena affecting human life, be carefully studied by strategists.   The northern Russians are not more hardy than the southern Turks, but it would have been madness for any Grand Vizier to have advanced far into Russia at a time of the year when heavy snow would close up all his lines of communication.

Perhaps in no part of Europe is the influence of climate more apparent than in Germany.   The people of northern, central, and southern Germany can be easily discriminated—far more easily, indeed, than the people of Yorkshire and of Devon—not only by dialectic peculiarities, but by habits and aptitudes.   The northern

Germans, dwelling by the mouths of the Ems, Weser, Elbe, Oder, and Vistula, have had a hard struggle with nature for ages in clearing and cultivating the marshy, cold, and densely-wooded territories in which their trying lot has been cast, and it seems but natural that these people should have a self-contained and resolute temperament. The climatic conditions of southern Germany, on the other hand, tend to an open air existence, to gaiety, and song. Central Germany is picturesque, and has been the home of brilliant literary genius. Jena, Weimar, Leipzig, and Gotha have long been centres of intellectual activity. Moreover the arrangements of the valleys, separated from each other by mountain walls, have favoured the development of the local spirit, and explain the partition of this country in feudal times into numerous principalities, of which some still exist.

Under the defective hygienic conditions that prevailed some fifty years ago, being ordered on service up the rivers of West Africa was almost equivalent to a sentence of death. The fever-stricken Campagna has been the grave of many a brave soldier when marching on Rome; Hannibal, for instance, suffering much in his own person. In countries with sandy surfaces the nights will probably be comparatively very cold, although in the day the highest temperatures prevail. The mean summer temperature of the deserts of Arabia, Persia, Africa, and the Punjab is 95°. The effect of forests is to make the days cooler and the nights warmer. Continental districts are much cooler than districts in the same latitude which are deeply indented by the sea. The south-west wind prevails in Great Britain, and is moist, being oceanic and equatorial, while the north-easter with us is dry and parching, being arctic and continental. By all reasons of latitude, the mean winter temperature of London should be 17°, but actually it is 38°. The mean winter temperature of the southern States of North America is almost the same as that of Lower Egypt, where there is seldom frost, and violent alterations of temperature are rare. The States by the Gulf of Mexico, however, are liable to both these evils ; in southern Texas with a violent norther the ther-mometer has fallen from 81° to 18° in 41 hours. The territory from Alaska to Lower California, by reason of the prevailing winds and the contours of the mountains, presents more variety of temperature than any other part of the world,—while the United States east of the Mississippi afford admirable climatic conditions for all varieties of trees, cereals, and grasses.

The people of the pampas of South America have had as yet no campaigns that need be chronicled, but those of the steppes have left indelible marks on the pages of history. What are now mere deserts were once the seats of mighty dynasties. Comparing the Asiatic steppes with the pampas of South America Humboldt says :—
" In that part of the Steppes inhabited by the Kirghiz and the Kalmucks which I have traversed, that is to say, from the Don, the Caspian Sea, and the Ural, to the Obi and the Upper Irtysh near the Dsiasang, over a space of forty degrees of longitude, one can never discover, even at the most distant limit, a phenomenon frequent in the llanos, the pampas, and the prairies of America—that horizon vague and boundless as the sea, which seems to support the vault of heaven. Seldom in

Asia was the spectacle offered me of even a single side of the horizon. The Steppes are traversed by numerous chains of hills, or covered with forests of conifers.

"The dust is whirled off the ground by the wind, and swept about in revolving tornadoes. The Steppes situated in a comparatively low latitude thus alternately assume the most discordant aspects. In winter the heavy rains inundate them, and transform them into impracticable marshes, spring clothes them with a thick carpet of grasses and other herbaceous plants, so that they reveal to the eye leagues upon leagues of delightful sward, cropped by numerous flocks. In summer they undergo a third metamorphosis, and are converted into parched and sun-scathed deserts like those of Nubia and Arabia.

"These periodical transformations are especially remarkable in the Steppes of the Black Sea, the Sea of Azov, and the Caspian Sea ; where winter comes attended with abundant snows and terrific tempests. No obstacle can arrest the fury of the gale, which accumulates the driven snow in fearful avalanches, and like the demon in the old German legend, drives before it the wild horses in an access of violence. The hurricanes are neither less numerous nor less furious in the hot than in the cold season ; dust, however, takes the place of snow, when, as is sometimes the case, no tremendous deluge of rain follows the track of the mighty wind. To sum up, the spring and summer of the Steppes are compressed, so to speak, into two months; all the rest of the year seems to be given over to desolation. Two months in the year of bloom, and sunshine, and colour, and beauty, are all that Nature grants the wandering Mongolian[1]."

[1] Humboldt, *Ansichten der Natur*, vol. I. Appendix.

In support of the historical examples adduced above
to show that military valour and success have not depended on climatic conditions the opinion of General
Marselli may here be quoted.   He says:—

"The careful observation of historical facts proves
that any nation, if moved by an imperious need or by
a noble idea, and if placed under a good disciplinary
system, can become a brave and warlike nation.

"Possibly we may find dispositions and capacities
more suitable for military services and warfare in mountainous countries and in temperate climates than in
monotonous plains and in districts of enervating heat ;
but history shows that the temperate zones at least
contain neither regions of courage nor regions of
cowardice.   A nation which to-day may be held by
general opinion to be unfit for war, may be to-morrow
rich in military characteristics, braced up by a virtuous
indignation and placed under a vigorous command ; on
the other hand, enervating moral influences can render
even the children of the mountains in temperate zones
indolent and feeble.   Nature predisposes character, but
she does not firmly decide it.   When historical conditions
permit it, all men can be brave soldiers[1]."

At 9800 feet or more above the sea-level the rarefica-
tion of the air produces in many people, and especially in
those unused to mountains, the disorder called mountain
sickness.   Even at lower altitudes, say 7000 feet, in the
Alps there are great differences of temperature between
day and night.   The night temperature is generally so
low in summer that soldiers unaccustomed to breathe in
the mountain air cannot bivouac in the open.   Above

---

[1] *La Guerra e la sua Storia*, vol. II., p. 155.

these altitudes, even Alpine troops cannot sleep in the open air except, perhaps, in well-sheltered spots, with plenty of fuel at their disposal. In these localities fuel has to be procured beforehand, as houses are few and small, and there is no arboreal vegetation.

Agriculture has almost as great an influence on Strategic Geography as the original natural state of the country. Well-cultivated districts are thickly populated, hence the towns and hamlets are close, and obstacles to the free movement of troops are multiplied ; on the other hand forests are thinned, scrub is removed, and swamps are drained, and thus the operations of farmers facilitate those of armies.

There have been many cases in modern wars in which mere accidents of climate and weather have been disastrous to armies, but in most of these over-confidence or official negligence has done quite as much harm as natural forces. For example, Napoleon's invasion of Russia was planned with reckless lack of prevision ; whether advancing or retreating, the march of each corps was conducted without reference to those who followed on the same road. It was not the snow only that ruined the Grand Army, its effectives were reduced one half before the retreat began, and the bad weather simply hastened what would have been a disaster in any case. Before a week had passed, and long before summer was over, 10,000 horses had perished for lack of nutriment.

Napoleon was rash ; and in war Fortune has only crowned with success those whose daring—like his in all previous campaigns—had no relation to caprice, whose initiative was based on elaborate and profound calculations, and whose careful preparation left nothing

to chance. If his counsels to his brother in Spain in
1812–1813 be compared with his own proceedings as he
marched to the Niemen and the Dnieper it will be seen
that in the former year he was predestined to disaster.
Yet when he recovered himself in 1813, all Europe in
arms could with difficulty force him and his raw levies
back to the Rhine.

That snow is a formidable obstacle to military progress can be learned from many a campaign, but on some
occasions ice has been an avenue to success. It was
over the ice that the troopers of Pichegru captured the
Dutch fleet on the Texel in 1795. Barclay de Tolly
led the Russians over the frozen Gulf of Bothnia to
Sweden in 1809, and the ice-bound condition of the river
St Lawrence in winter complicates the question of the
defence of Montreal and Quebec against the United
States.

Frost was fatal even in the Peninsula, and impru-
dence in the use of alcohol increased the dangers from
bad weather. From these causes, in January, 1813, no
fewer than 150 of King Joseph's French guards were
frozen to death in the Guadarrama Pass. Wellington's
movements on the Agueda during the same month were
hampered by snow. Yet Napoleon in his eagerness to
cut off Moore had crossed the Sierra Guadarrama with
but little loss in December, 1808, when it was covered
with snow.

The remarks of Quintus Curtius about the sufferings
of Alexander's men in the Caucasus may be applied
word for word to the retreating French in Russia after
November 6th, 1812. " Dreary scenery and impassable wilds terrified the exhausted soldiers, they were

astonished by solitudes without a vestige of cultivation or of man. So deep were the snows that shrouded the ground, which was bound fast by ice and frost, that no sign was perceived of birds or any beasts remaining out. The light was rather an obscuration of the sky resembling darkness."

But it was reserved for the Russians themselves to afford the latest example of the terrors of mountain passes when frost and snow set in. At Shipka their 24th Division lost over 6000 men during the storm of the 18th to 23rd December, 1877. Gourko lost 2000 men frozen to death during this same storm. Again, in the movement from Plevna to Philippopolis, Daudeville lost 1000 more; while during the march to the valley of the Maritza bad food and the lack of change of clothing laid the seeds of typhus and typhoid, which soon broke out with terrible malignity.

A sudden thaw is frequently as adverse to the progress of armies as a snowfall. During their movement from St Dizier to Brienne in January, 1814, the French troops underwent the most dreadful fatigue in forcing their way through the deep and miry alleys of the great forest of Der. The frost had given way, and the thaw which succeeded had rendered the execrable cross-roads all but impassable. It was only by the greatest efforts that the guns and artillery waggons could be dragged through; the peasants harnessed themselves to the guns and toiled night and day through the mud till at length the forest was passed and the exhausted troops emerged into the open country. Throughout the whole campaign of 1814 in the deep and heavy soil of Champagne, with bad country roads and wretched weather, the sufferings of the soldiers on both sides, who were constantly manœuvring, were aggravated to an almost intolerable degree[1].

How fallings out and mortality may be diminished in the same climate is strikingly exemplified by a comparison of the operations of Wellington in the autumn of 1812 with those of the spring of the following year. During the march from Salamanca to Madrid in 1812, two men in ten fell to the rear, while during the march from the Douro to Burgos in 1813 not more than 8 in 500 dropped on the march. No better proof could be afforded of the excellent management of the Commander-in-Chief, when it is remembered that there were not less than 80,000 men moving forward in the same direction at the same time within touch of each other, with cavalry, artillery, tents, and baggage.

Climate remaining the same, mortality varies with hygienic arrangements. The loss in our West Indian battalions has diminished in a most gratifying manner in a few generations, but the Russians in Turkey in 1878 were in nearly as bad a state as their predecessors in 1829.

Everything went wrong with the unfortunate Walcheren Expedition in 1809. When already half spoiled by naval, military, and official incompetence it was ruined by fever. The disease first showed itself amongst the troops in South Beveland who had not the opposition of an enemy to keep their minds and bodies in healthy action, but after the fall of Flushing it broke out among the troops in Walcheren. The island, being so flat, is little better than a swamp; the ditches are filled

[1] See Alison, *History of Europe*, chap. lxxxv.

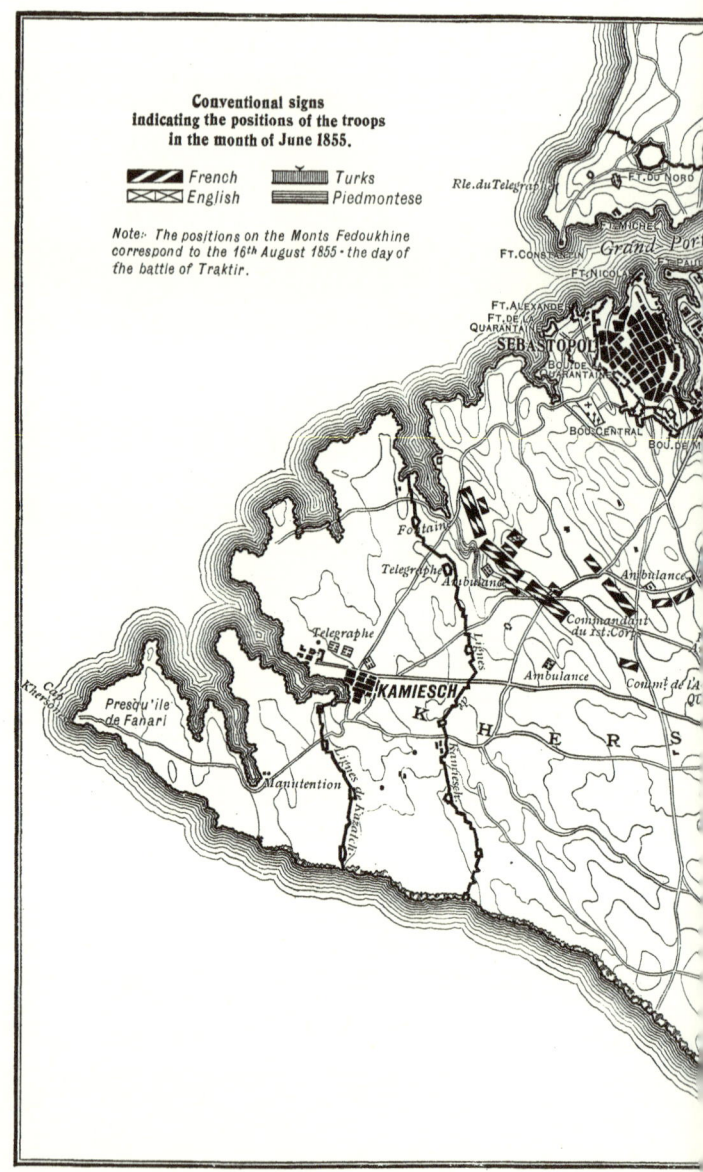

Conventional signs indicating the positions of the troops in the month of June 1855.

French
English
Turks
Piedmontese

Note:- The positions on the Monts Fedoukhine correspond to the 16th August 1855 - the day of the battle of Traktir.

Rle. du Telegr...
FT. 60 NORD
FT. MICHEL
Grand Port
FT. CONST...
FT. NICO...
FT. ALEXANDER
FT. DE
QUARANTA
SEBASTOPOL
BOU. DE
QUARANTAIN
BOU. CENTRAL
BOU. DE M

Fontai...
Telegraphe
Ambulance
Ambulance
Commandant
du 1st. Corps
Telegraphe
Ambulance
Comm.t de l'A
KAMIESCH
K
H
E
R
S
Cap. Kherso
Presqu'île de Fanari
Manutention
Lignes d'Aug...
Lignes de Kam...

MAP OF SEBASTOPOL AND ITS ENVIRONS IN 1855.

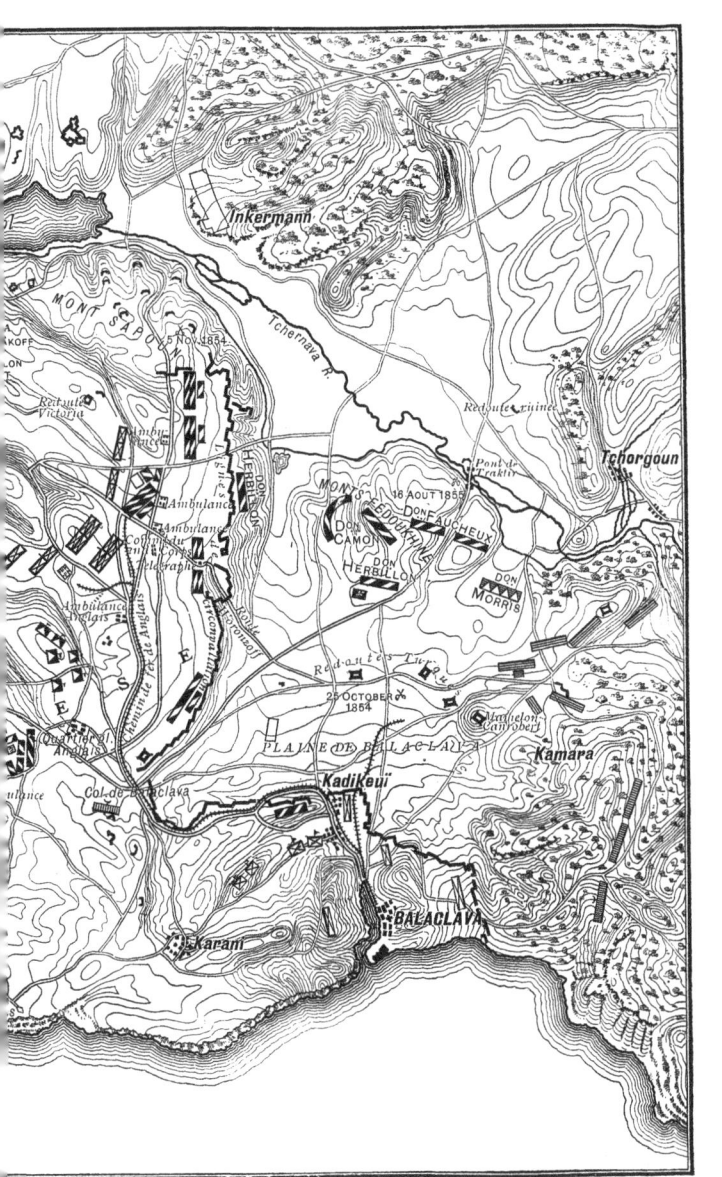

Inkermann

MONT SAPOU 5 Nov. 1854

Tchernaya R.

Redoute ruinée

Redoute
Victoria

Pont de
Traktir

Tchorgoun

Kamilly
Inkerm

MONT S 16 AOUT 1855
DON PÉLISSIER
Ambulance DON
CAMON DON
Corps du DON HERBILLON
télégraphe Ambulance MORRIS
en 1er corps
télégraphe

Ambulance
Anglais

Redoutes Turques
Quartier 25 OCTOBER
Anglais 1854 Mamelon
Canrobert

bulance PLAINE DE BALACLAVA Kamara

Col de Balaclava Kadikeui

Karani BALACLAVA

with putrid vegetable and animal matter, the quantity
of pure water is very limited. Nearly one-third of the
native population is regularly attacked by fever every
sickly season, in spite of their attention to cleanliness
in buildings and person, and no remedy could be devised
to check its ravages in the army. Even those who
recovered from the disease itself had their constitutions
so shattered that their physical power was materially
diminished. In July, 1809, the 81st Regiment had
650 men fit for duty, in September only 40, and of
35,000 officers and men who returned to England
11,000 were in hospital.

During one of Grant's campaigns in Virginia the
weather had been fair for several days, and the roads
were getting in as good condition for the movements of
the troops as could be expected, for in that section of
the country, in summer, the dust was usually so thick
that the army could not see where to move, and in
winter the mud was so deep that it could not move
anywhere.

"The weather now began to get cloudy, and towards
evening rain began to fall. It descended in torrents
all the night, and continued with but little interruption
during the next day. The country was densely wooded
and the ground swampy, and by the evening whole
fields had become beds of quicksand, in which the
troops waded in mud above their ankles, horses sank
to their bellies, and waggons threatened to disappear
altogether. The men began to feel that if anyone in
after years should ask them if they had been through
Virginia, they could say, 'Yes, in a number of places.'
The roads soon became sheets of water, and it looked

as if the saving of that army would require the services, not of a Grant, but of a Noah[1]."

Out of 24,000 British who perished in the Crimea, only 4,000 were wounded ; the remainder died of cholera and other diseases brought on by hardship and exposure, and no small proportion of the deaths were due to neglect.

In Cuba, in the war of 1898, about 600 United States soldiers died of wounds, but some 6000 or more of disease, and a large proportion of these deaths were due to defective medical and sanitary arrangements. This is how Mr Atkins describes the condition of the Army Medical Corps and the weather. Speaking of the night after the fighting of 3rd July, he says :—

" There were not nearly enough tents, cots, medicines, doctors, nurses, or carriers. Everything was insufficient. I have never seen anything more pitiable than the spectacle of wounded men lying all night without a tent-covering over them on the muddy ground and in the soaking dew. Night on a hospital ground was a time of horror ; there was moaning everywhere, and one night I remember two men calling all night for someone to kill them."

Of another night, over a week later, he says :—

" A thunderstorm came—such thunder as I have never heard and never thought to hear—so near, tremendous, and splitting. With it came a tropical storm of rain, falling in a wall so that you could not see through it. Soon the ground where I lay was under water. A volunteer regiment had arrived late at night,

[1] Porter's *Campaigns with Grant.*

and had no time to encamp themselves; the morning revealed them lying in a lake. The horses were all frightened with the storm, and came round the tents whinnying. And in the middle of it all, two men who had been crying out deliriously in the 'hospital' began to wander about in the field gibbering. This was a hospital in which there were cases of yellow fever."

In support of the contention that foresight, plenty of clothing, abundance of food, good discipline, and good boots will enable healthy men to pass with comparative impunity over mountains covered by snow and ice, may be adduced Manteuffel's rapid march with two corps and 168 guns over the Côte D'Or, between Dijon and Langres, January 13th to 16th, 1871. At the same period Bourbaki's army was perishing without any. rapid marches in easy country, and he wrote to De Freycinet, "Men and horses are broken down with fatigue; you have no idea of the sufferings which the army has endured since the beginning of December. It is perfect martyrdom to hold its command." The brave old soldier broke down and tried to commit suicide, and his army fled into Switzerland, where a convention was signed between General Clinchamp and the Swiss General Hertzog, who with a large force had been guarding the neutral line. On 1st of February the relics of what had once been an army of 133,000 men crossed that line and laid down their arms. The Germans had captured about 15,000 men with 19 guns, before their escape to neutral territory could be effected; while 84,000 surrendered to the Swiss. Most of these unfortunate men—surely the most to be pitied of any of the victims of the war—arrived in Switzerland in a state which defies description. "Their clothes were rent, and dropping off them in tatters; their feet and hands were frost-bitten. While the shrunk features and crouching gait told of gnawing hunger, the deep cough and hoarse voice bore witness to long nights spent on snow and frozen ground. Some had bits of wood under their bare feet to protect them from the stones; others wore wooden sabots; hundreds had merely thin cotton socks, and many none at all; others who appeared well shod would show a boot without sole or heel—the exposed part of the foot, once frozen, now presenting a wound crusted with dirt. For weeks none had washed or changed their clothes, or put off their boots. Their hands were blacker than any African's. Some had lost their toes; the limbs of others were so frozen that every movement was agony. The men stated that for three days they had neither food nor fodder served out to them, and that even prior to that period of absolute famine one loaf was often shared between eight of them[1]."

In his *Defence of Plevna* Capt. W. V. Herbert describes the horrors resulting from severe climate and scarcity of food :—

" The sentry service in our own redoubt, as well as throughout camp, was of a cruelly severe character in the rigour of a Bulgarian winter. The original four hours had to be reduced to two, then to one hour. Fixed, almost buried alive, in a hole four feet deep, with the upper part of the body exposed to the bitter blasts, the lower embedded in the frozen ground, unable to move (the slightest attempt at a trot, the very act of stepping out of the hole, attracted the enemy's bullets),

[1] Colonel Hozier, *The Franco-German War*, II. p. 250.

insufficiently fed, compelled to exercise a ceaseless vigil-
ance, struggling against the dangerous drowsiness engendered by frost, the men looked upon sentry-duty as
the last refinement of torture.   Our splendid great-coats
were invaluable to us.   When snow was on the ground
the cold was less severely felt : snow with five degrees
below freezing-point was better than one or two degrees
above freezing-point without snow.   The long, winding
line of sentries, lost in the murky distance of a bleak
winter day, with only the dark hoods and the bayonets
visible on the white ground, presented a grotesque and
striking appearance.

"By the beginning of November the rations had
already been reduced, more particularly as regards
meat. Bread made of maize-meal, and baked in Plevna,
took the place of biscuits, the large stock of the latter
commodity being retained in view of a possible sortie
and a march across a famine-stricken country."

Another quotation from the same work will illustrate
the awful scenes which result from defective medical
arrangements :—

"There was a deficiency of drugs, quinine was almost
entirely absent.   Lint was wanting, garments had to be
cut up for bandages, however much clothing of every
description was in demand ; wounds could not be bound
up during the last days of the investment for want of
material.   The convalescents had no strengthening food.
Invalids quarrelled for precedence. The German surgeon
Lange said that he had not taken off his clothes for
four weeks and had no more than three hours' sleep
per night."

This was in Plevna in 1877 : similar scenes took place

in Beaugency in France in December 1870. In the
theatre alone were upwards of 200 desperately wounded
men. For many hours there was no medical man in
the place. As the wind was intensely cold, diminished
circulation hastened the end of many a man who would
have been easily cured if attendance could have been
prompt. The dead and dying lay close ; as the former
were removed their places were forthwith filled. Even
water, for which there were incessant demands, could not
be procured in sufficient quantities.

During Chanzy's retreat, December 1870, the weather
had been dreadful. On December 12th it was particularly
bad. A torrent of rain had melted the snow and produced a thaw. The roads were everywhere exceedingly
slippery, and the fields were too muddy for the passage
of horses and carriages. Nevertheless the march was
effected with a reasonable degree of regularity.

But, severe as were the sufferings of the retreating
French, the pursuing Germans were in nearly as bad a
case. On January 9th the roads were once more as
hard as iron from frost, and were covered with ice, which
remained for days and made the cavalry nearly useless
in the actions round Le Mans. The Commander-in-
chief had to dismount and walk, his staff were in the
same plight unless they tried to ride in the ditch by the
side of the road. The artillery and train horses were frequently falling, still the army was compelled to press on.

Cold in itself is not very trying to healthy men.
Nansen and his comrades enjoyed the very best of health
during their polar expedition, and it will be remembered
that he and Johansen actually gained in weight during
the sledge-journey. Soldiers will stand cold well if well

fed, well clad, and provided with warm shelter, as were Von Werder's men in 1871, but to the troops of Prince Frederic Charles marching on Le Mans at the same time, the slippery condition of the roads and the severity of the weather were severe hardships, and the difficulties were increased by fogs and mists in a close country.

Captain H. H. Deasy, late of the 16th Lancers, reached Yarkand on the 2nd February, 1899, after a three months' winter exploration along the valley of the Yarkand river and the adjacent country, from the western end of Rashkam at the foot of the Karakoram mountains to near Yarkand. The greater part of the route taken led through country which had never before been traversed by any European. Numerous steep and difficult passes, only crossed by execrable tracks, retarded his journey. To survey one stretch of the Yarkand river about eleven miles in length, a detour of nine marches had to be made, five passes crossed in mid-winter, one of them 17,000 feet in height, and five nights spent out in the open when the average minimum temperature was twenty-seven degrees below zero.

Some countries are turned into quagmires by even a day's steady rain ; thus the routes in Belgium from Ligny to Gembloux, from Quatre Bras to Waterloo, and from Wavre to Frichermont were very difficult by reason of the rain of June 17th, 1815, and this fact had a most serious effect on the plan of Grouchy, Napoleon, and Blucher. The officials and the mass of the people at the capitals frequently do not take these impediments into account, and generals are censured for enforced inaction. The Washington authorities were indignant at McClellan's slowness in 1862, but they had not the General's experience of Virginia mud. On two occasions during his Yorktown campaigns, "the divisions of Franklin, South, and Porter were with difficulty moved to White-house, five miles in advance; so bad was the road that the train of these divisions required thirty-six hours to pass over this short distance." Again,—"The supply trains had been forced out of the roads to allow the troops and artillery to pass to the front, and the roads were now in such a state, after thirty-six hours con-tinuous rain, that it was almost impossible to pass empty waggons over them[1]."

The great scourge of armies in the past has been dysentery, and any but the hardiest troops succumb in great numbers to the consequences of lying on damp ground. In this respect both besiegers and besieged were particularly unfortunate at Metz in the autumn of 1870. The French troops were bivouacked outside the town, between, and outside, the forts. They were insufficiently fed, their inertia had a bad moral effect which reacted on their health, but above all, the heavy rains in September and October made their state cheer-less and unhealthy to a degree, and they were heavily smitten with sickness. The German investing soldiers had a more hopeful prospect; they were elated with victory and were well fed, but sturdy as the men were, remaining stationary so long on great battle-fields or charnel fields soon told upon them. In some divisions 50 per cent. were ill.

The French, too, suffered severely in the Madagascar campaign. Mr Bennet Burleigh gives a sad picture of the combined effect of climate and incompetence on their

[1] Hamley, *Operations of War*, p. 21.

vitality. The mischances and mistakes were endless. Owing to the labour of constructing roads, as well as other causes, the sickness and mortality were heart-rending. Out of some 15,000 men 6000 at least died from the effects of climate, and of wounds only 21. Not a man among the troops escaped fever. In one of the transports, the Ville de Metz, conveying sick back to France, there were 93 deaths on the voyage.

Compared with this record, the experiences of the British in their recent expeditions in Africa, north, south and west, have been most reassuring. Sir Charles Wilson speaks in the most favourable terms of the climate from Korti to Khartum, and Colonels Baden Powell and Alderson are enthusiastic about the delights of outdoor existence in Matabililand and Mashonaland. With re-gard to supply, transport, and sanitation during our recent operations from Chitral to Tirah, and from Benin and Ilorin to the Egyptian Sudan, our troops seem on the whole to have been better managed in all these striking diversities of climate than has ever been the case before with the warriors of any world power except the Romans at their best.

The utmost difficulties arising from variations in temperature, climate, and character of the country have been met and overcome by British troops; in frequent instances the same battalion has in the same year endured the utmost caprices of nature. To illus-trate the various climatic experiences of our regiments, examples under the territorial system can be supplied by one English, one Irish, and one Scotch regiment. The Liverpool Regiment in this century has fought in such different localities as Martinique, Niagara, Delhi,

Peiwar Kotal, and Burma. The Royal Irish Regiment has fought in the same century at Pegu, Sebastopol, New Zealand, Afghanistan, Tel-el-Kebir, and the Sudan. The Royal Highlanders have fought in Kaffraria, in the Crimea, at Lucknow, in Ashanti, at Tel-el-Kebir, and in the Sudan.

The vicissitudes of climate, though often very trying, have not produced the least effect on the fighting vigour of our troops. Even when particular regiments were well-nigh decimated by sickness and other hardships, the men left fit for the battle were always true to their old traditions. As Sir Rennell Rodd says:—"Britain has never failed to find among her sons the men that she has need of. And they will never fail her till she turns her back on Empire, and forgets the sea."

INDEX.

For EU product safety concerns, contact us at Calle de José Abascal, 56–1°,
28003 Madrid, Spain or eugpsr@cambridge.org.

www.ingramcontent.com/pod-product-compliance
Ingram Content Group UK Ltd.
Pitfield, Milton Keynes, MK11 3LW, UK
UKHW010850090126
466816UK00011B/138